DARWIN'S DNA

This is an illustrated version of Dr. Andrea Diem-Lane's book, *Darwin's DNA* which has been republished in a small paperback under the title, *The DNA of Consciousness*.

MSAC Philosophy Group | 2016

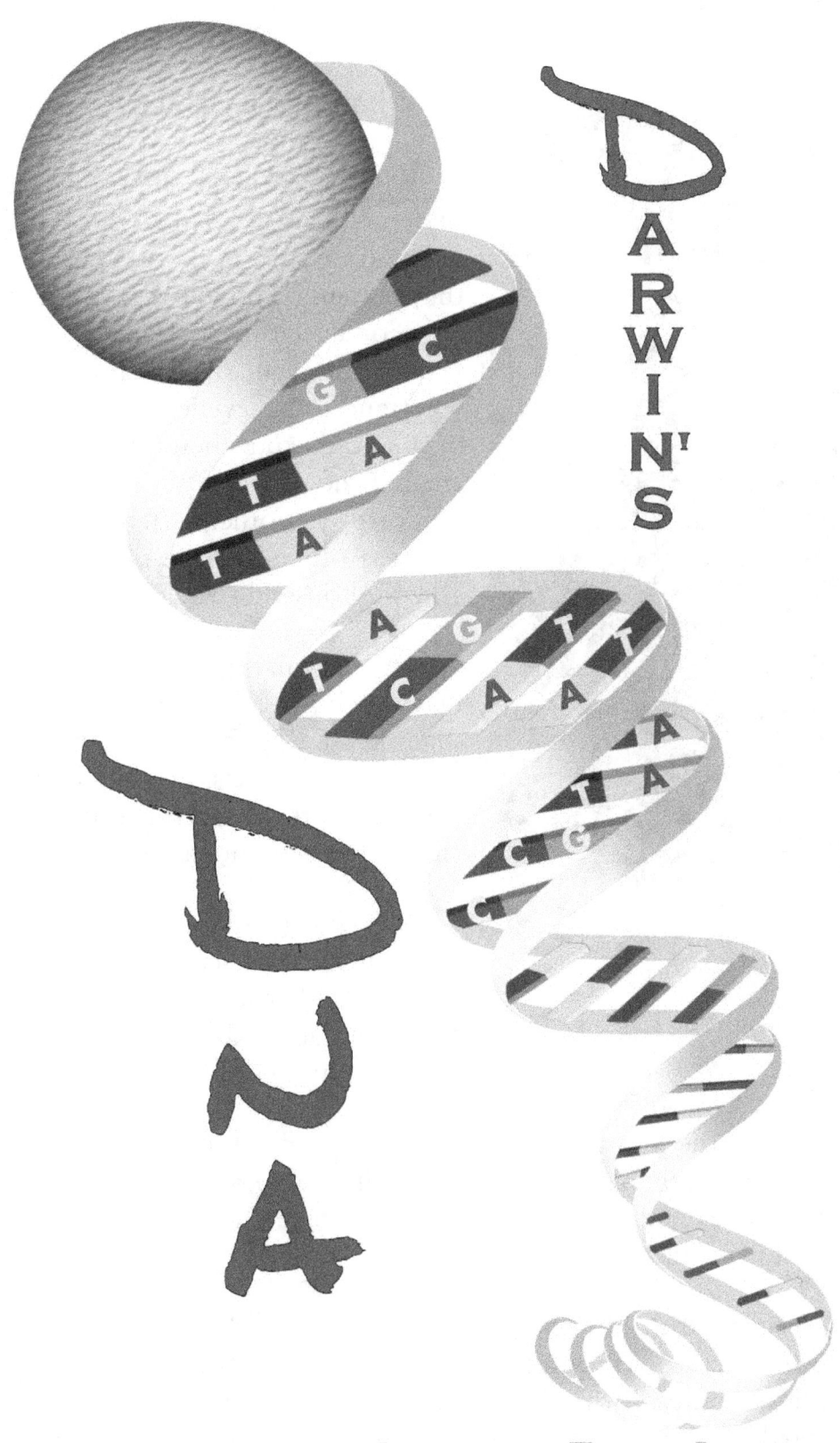

Explaining Evolution

One of the difficulties in explaining the theory of evolution is where to start. I have noticed over the years in teaching this subject at the undergraduate level that most students don't really know much about Charles Darwin and they certainly know less about the mechanism of natural selection. Usually, I have a fairly vocal contingent of fundamentalist Christians who tend to believe in the literal interpretation of Genesis and therefore believe that God created the world in six days. Some accept an old earth hypothesis (five billion years old); whereas others believe that the world is only several thousand years old.

Given this mixture in my class, my husband, David, who is also a Professor of Philosophy, suggested a way that not only reaches students of varying persuasions but ends up actually convincing them, even if only partially, of the viability of evolutionary theory.

Whenever you have to present a controversial idea, he has noticed it is best to begin with what we all know and what we all agree upon. Thus, he advised that I ask for a student volunteer and use him or her as my example.

Let's imagine that the young man's name is Shaun. He is 18 years old. The first question we pose is a simple one. Why does Shaun physically look the way he does? Excepting his clothes and hygiene and other personal choices, my students invariably say "because of his parents."

And that is certainly true and nobody tends to dispute the fact that his parents were the key to his physical looks. But what is it that his parents bring to the equation? The simple answer, of course, is that Shaun's father (let's call him Christopher) contributes sperm whereas his mother (let's call her Catherine) an egg. Both substances are quite small. A human sperm for instance (including both its head and its tail) is roughly 55

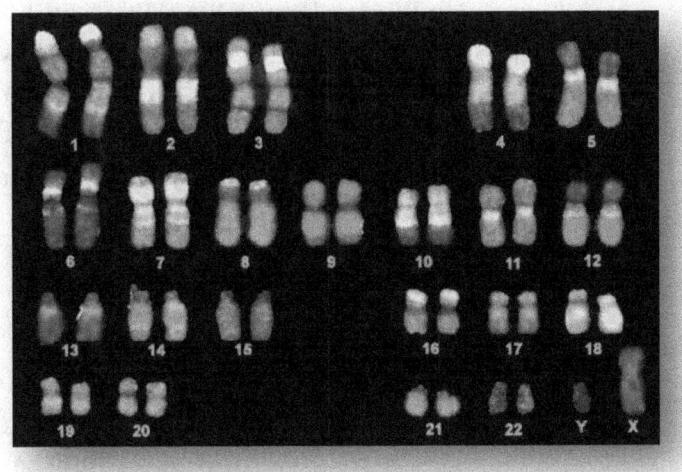 microns, so tiny that it is 25,000 times smaller than a golf ball. At this point, we start to see one of the first hallmarks of science and one that is often misunderstood: *reductionism*. The contribution that Shaun's parents made to their offspring doesn't come at the phenotype level (that is merely the replicating appliance), but rather at the genotype particulate. We have scaled down from a 6"1 body and a 5"4 body to precisely what those larger frames house—a sperm and an egg. Now we have literally "reduced" Shaun's parents to the arena where the information they share is more easily accessible and localized.

One metaphorical way of putting this is to imagine that Christopher and Catherine are individual books, filled with all sorts of historical information. Their desire is to recombine their books and produce a new edition. Since human sexual intercourse is literally the intertwining of two fairly large body forms so that a transmission of information can catalytically induce a viable recombination to occur, we can readily see that what makes a child is essentially the intersecting of binary forms of data.

The larger question that arises in this metaphor (with its real world applications) is how to decipher Catherine's and Christopher's contributions to the book they named Shaun. In what language are their books written? Are they the same language or different? What is the alphabet or rudimentary notation wherein their respective knowledge is inscribed?

Today, unlike in Darwin's day, we can actually answer these queries with remarkable accuracy. Although an egg and a sperm are compositely varied, the essential code they contain is written in a biological language known most famously by its initials: D.N.A. or more properly deoxyribonucleic acid. This language comes in four letters, easily remembered with the acronym C.T.A.G. Here "c" stands for cytosine, "t" for thymine, "a" for adenine, and "g" for guanine.

Thus, Catherine and Christopher shared books which had a common four lettered language. These letters further comprised whole pages numbering in the tens of thousands known as genes which formed twenty-three chapters known as chromosomes. Our entire book is known as a genome.

| Darwin's DNA

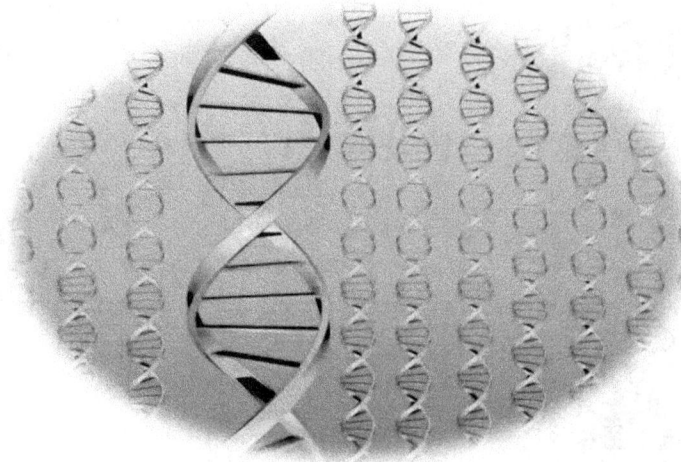

Shaun is, therefore, the result of two genomes (books) recombined which results in a unified outcome of 25,000 plus genes (pages), 23 chromosomes (chapters), all written in DNA (a four letter alphabet).

Sexual selection, however, is merely the start of why Shaun looks the way he does. Since obviously his mother could have chosen another author to help write her memoirs and in so doing produce a completely different son with a unique set of genes. Shaun isn't merely a duplicate reconfiguring of his parent's deoxyribonucleic acid. Occasionally, when DNA is copied throughout one's body it makes a copying error, what is known as a mutation. This can simply be a one letter change. Instead of ATGGTTTGATGTC, one might get ATGGTTGATTGTC. While at the level of just script it looks somewhat inconsequential, but in terms of applied genetics such minor variations can loom large.

Why mutations occur is a deep subject, but not an insurmountable one, especially if you have an understanding of Heisenberg's uncertainty principle and how quantum mechanics is based on indeterminacy. The introduction of "chance" (or occasional mutations) into the genome is important because it causally explains how a child can indeed be unexpectedly and unpredictably different than his parents. In other words, sexual selection is just one component in what makes up Shaun's physical characteristics. DNA mutations are another. There are two fundamental ways to mutate a gene. The first way is through environmental damage. As the *Learn Genetics* website, sponsored by Genetic Science Learning Center at the University of Utah, explains:

Ultraviolet light, nuclear radiation, and certain chemicals can damage DNA by altering nucleotide bases so that they look like other nucleotide bases.

When the DNA strands are separated and copied, the altered base will pair with an incorrect base and cause a mutation. In the example below a "modified" G now pairs with T, instead of forming a normal pair with C.

Environmental agents such as nuclear radiation can damage DNA by breaking the bonds between oxygen (O) and phosphate groups (P).

| Darwin's DNA

Cells with broken DNA will attempt to fix the broken ends by joining these free ends to other pieces of DNA within the cell. This creates a type of mutation called "translocation." If a translocation breakpoint occurs within or near a gene, that gene's function may be affected.

The second way is by DNA replication. As the same website elaborates:

Prior to cell division, each cell must duplicate its entire DNA sequence. This process is called DNA replication.

DNA replication begins when a protein called DNA helicase separates the DNA molecule into two strands.

Next, a protein called DNA polymerase copies each strand of DNA to create two double-stranded DNA molecules

Mutations result when the DNA polymerase makes a mistake, which happens about once every 100,000,000 bases.

Actually, the number of mistakes that remain incorporated into the DNA is even lower than this because cells contain special DNA repair proteins that fix many of the mistakes in the DNA that are caused by mutagens. The repair proteins see which nucleotides are paired incorrectly, and then change the wrong base to the right one.

At this junction, most of the class is still with me, since they can readily concede that sexual selection and chance play a huge role in producing children. What is not so apparent is how a four lettered language could produce the wide array of complexity that we see in the human species. In other words, how can such diversity arise from only four varying molecule clusters?

To explain it better, I start again with what we know and then posit a query. The English language has 26 letters and from that can we get a wide diversity of published materials? The answer is an easy yes. Just go to the local library and you can see what diversity such a language can bring—ranging from Cosmopolitan magazine to US weekly to the New York Review of Books to Surfer's Journal. All telling different stories, yet all written in the same English alphabet utilizing A's to Z's. But what would happen if we only had 2 letters, not 26; could we still have the same wide diversity? Usually, at this moment, my students shake their heads with a face of disapproval, wrongly imagining that more letters would mean greater diversity.

Such is not the case, however. If we only had two letters or two numbers (0 and 1), we could, in point of fact, reproduce our entire alphabet and our entire English set of words, sentences, paragraphs, and books. Indeed, we could recapitulate any system of information that has appeared on earth. The very basis of computer programming is predicated upon a string of 0's and 1's. The internet itself can be seen as a huge ocean of cascading binary bits, triggering off an electron dance near the speed of light where information reaches innumerable portals.

The mythic universal library, so hauntingly brought to life in Jorge Borges' famous short story, "The Library of Babel," which was written before the advent of the World Wide Web, indicates an almost infinite range of information. W.V. Quine, the late professor of philosophy at Harvard University, ironically pointed out that letters were actually unnecessary, since even a dot and a dash could suffice and all the texts of the world could be replicated by such a simple binary. As Quine summarizes, "The ultimate absurdity is now staring us in the face: a universal library of two volumes, one containing a single dot and the other a dash. Persistent repetition and alternation of the two is sufficient, we well know, for spelling out any and every truth. The miracle of the finite but universal library is a mere inflation of the miracle of binary notation: everything worth saying, and everything else as well, can be said with two characters. It is a letdown befitting the Wizard of Oz, but it has been a boon to computers."

Even here, one could argue that Quine didn't go far enough, since all one would need is a dot and its absence. The dash itself being unnecessary since a thing and its absence is sufficient to be its own binary system.

Thus, it comes as a surprise to my students to realize how easy it is to get such wide diversity from a simple language or code. But even though sexual selection and genetic mutations can explain much, they are not sufficient to explain a much larger process called natural selection, for which Charles Darwin is rightly famous.

| Darwin's DNA

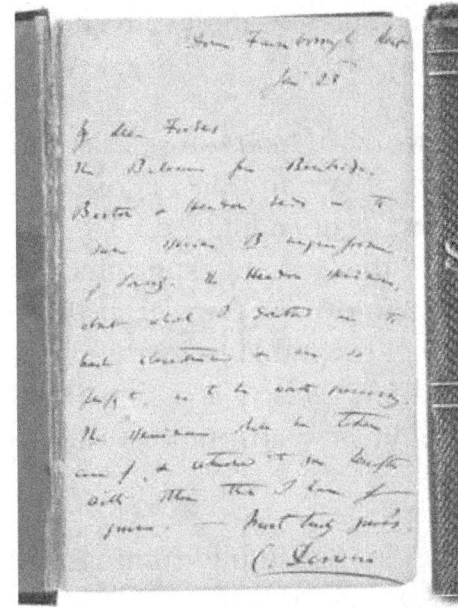 

Natural selection, as defined by Darwin in his famous *On Origin of Species*, is:

> Owing to this struggle for life, any variation, however slight and from whatever cause proceeding, if it be in any degree profitable to an individual of any species, in its infinitely complex relations to other organic beings and to external nature, will tend to the preservation of that individual, and will generally be inherited by its offspring. The offspring, also, will thus have a better chance of surviving, for, of the many individuals of any species which are periodically born, but a small number can survive. I have called this principle, by which each slight variation, if useful, is preserved, by the term of Natural Selection, in order to mark its relation to man's power of selection. We have seen that man by selection can certainly produce great results, and can adapt organic beings to his own uses, through the accumulation of slight but useful variations, given to him by the hand of Nature. But Natural Selection, as we shall hereafter see, is a power incessantly ready for action, and is as immeasurably superior to man's feeble efforts, as the works of Nature are to those of Art.

Natural selection, though precisely defined by Darwin, has led to some unnecessary confusion since it mistakenly implies a conscious act on the part of nature to pick and choose. Rather, it may be more properly understood as *natural elimination* and anything that can survive that global and unending process is, because of such survival, sufficient (not necessarily "best") to continue on. Viewed in this purview, evolution by natural selection isn't so much about "fittest" or "strongest" or "best," but rather as contingently successful. With the operative word here being contingent since what is viable in one environmental niche may not be so in another.

What cannot be denied are the vast odds against life to survive under such harried conditions. That anything does survive tells us much about both the environment from where it arose and the competition it had to go head to head against in order to live long enough to pass on its code. Natural selection or natural elimination or survival of the sufficient (however, we describe this winnowing process) is fundamentally a description of how organic life struggles for a temporary respite from ultimate annihilation. We can witness this struggle right now in the world we live in. We hear of an earthquake in China, where thousands are summarily killed, and yet several individuals, defying astronomical odds, survive. We hear of powerful and relatively new viruses, such as HIV, which can kill millions after years of incubation.

And, yet, there are a few who seem to ward off its terminal sentence and live relatively long and healthy lives, even without the introduction of new drugs.

How is this possible? Variation. Natural selection only works if there are variations among organic life, where a panoply of potential body types live and die. Those that do survive this gauntlet (and the gauntlet, lest we forget, is unending), do so only temporarily and only under certain conditions. It is under this severe testing that we can start to see how certain adaptations are better suited than others. Further, if those adaptations can produce viable offspring that carry on such favorable traits then they will have a built-in advantage over competitors that do not. This isn't a static sort of testing, however, since environments change over time and new adaptations due to wide variability arise and compete anew.

As Darwin so beautifully summarized nearly 150 years ago:

It is interesting to contemplate an entangled bank, clothed with many plants of many kinds, with birds singing on the bushes, with various insects flitting about, and with worms crawling through the damp earth, and to reflect that these elaborately constructed forms, so different from each other, and dependent on each other in so complex a manner, have all been produced by laws acting around us. These laws, taken in the largest sense, being Growth with Reproduction; inheritance which is almost implied by reproduction; Variability from the indirect and direct action of the external conditions of life, and from use and disuse; a Ratio of Increase so high as to lead to a Struggle for Life, and as a consequence to Natural Selection, entailing Divergence of Character and the Extinction of less-improved forms. Thus, from the war of nature, from famine and death, the most exalted object which we are capable of conceiving, namely, the production of the higher animals, directly follows. There is grandeur in this view of life, with its several powers, having been originally breathed into a few forms or into one; and that, whilst this planet has gone cycling on according to the fixed law of gravity, from so simple a beginning endless forms most beautiful and most wonderful have been, and are being, evolved.

| Darwin's DNA

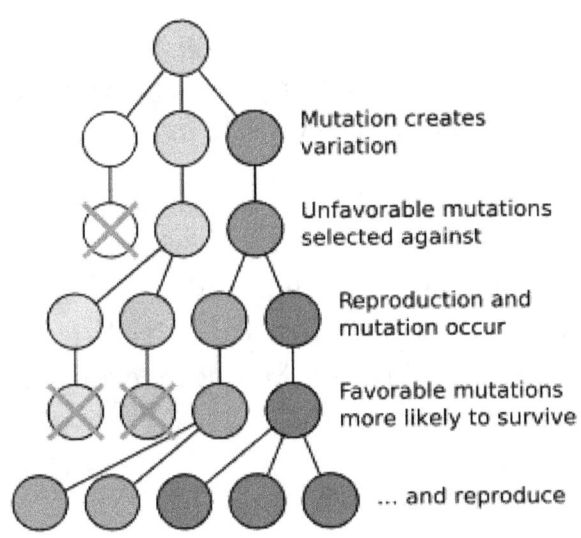

Now going back to Shaun, we have realized three key factors which have determined why he looks the way he does: sexual selection, genetic mutation, and natural selection. Play this out in your own life. Right now you are being tested by nature, even if completely unknowingly. Either you are going to have offspring that carry on your unique DNA configuration or you are not. If the former, you are in competition with other likeminded brethren in pursuing a potential mate. If the latter, you have basically given a Trumanesque-like gesture to millions of years of competitive successes that led up to your very being. You have said, "The buck stops here." And, barring some future resurrection of your DNA in unseen ways, the book that was you will go into the non-circulating library of forgotten achievers.

But sexual success, while necessary, doesn't insure lineal success, since as we have pointed out a random genetic mutation could mean that your child won't live long enough to procreate. And, even if your child does live long enough there are many other factors to consider, including new environmental conditions (such as global warming), new strains of virulent bacteria and viruses, and any host of unpredictable variables that may arise in the unforeseen future—all of which can wipe out even the most successful of genetic programs.

All of this competition has led to a natural "editing" of what we see today. If something survives, you know a priori it has been edited or pruned by the very competition which led to its survival. *Knowing the conditions from where we originally arose is a central key to understanding why we survive as we do in the present.*

Shaun, therefore, is in the most literal sense a genome with a long history, one which was shaped by factors which we cannot access fully today, but which nevertheless give hints of its long sojourn on terra firma.

Every strand of DNA contains a unique history of its journey and what must have transpired to shape it into its present incarnation. Thus, Shaun's history doesn't start with parents, or with his grandparents, or with great grandparents, but rather goes thousands, nay millions, of years back in time.

It is right here in the lecture when I ask Shaun about his ancestry. Where were you born? Where are your parents from? Where did your ancestors come from?

Nobody is indigenous to America. We all came from somewhere else. In Shaun's case, we found out that he could trace his family lineage back to Ireland and Germany. But that also is not the final resting place for his DNA. We now know that all human beings living today trace their genetic history some 90,000 plus years back to the heartland of Africa. Shaun is in this sense (as we all are) African-American, even if his latter day sojourn may have resting places in European countries.

As the entry on human migration in the *MSN Encarta* puts it:

Early humans first migrated out of Africa into Asia probably between 2 million and 1.7 million years ago. They entered Europe somewhat later, generally within the past 1 million years. Species of modern humans populated many parts of the world much later. For instance, people first came to Australia probably within the past 60,000 years, and to the Americas within the past 35,000 years. The beginnings of agriculture and the rise of the first civilizations occurred within the past 10,000 years.

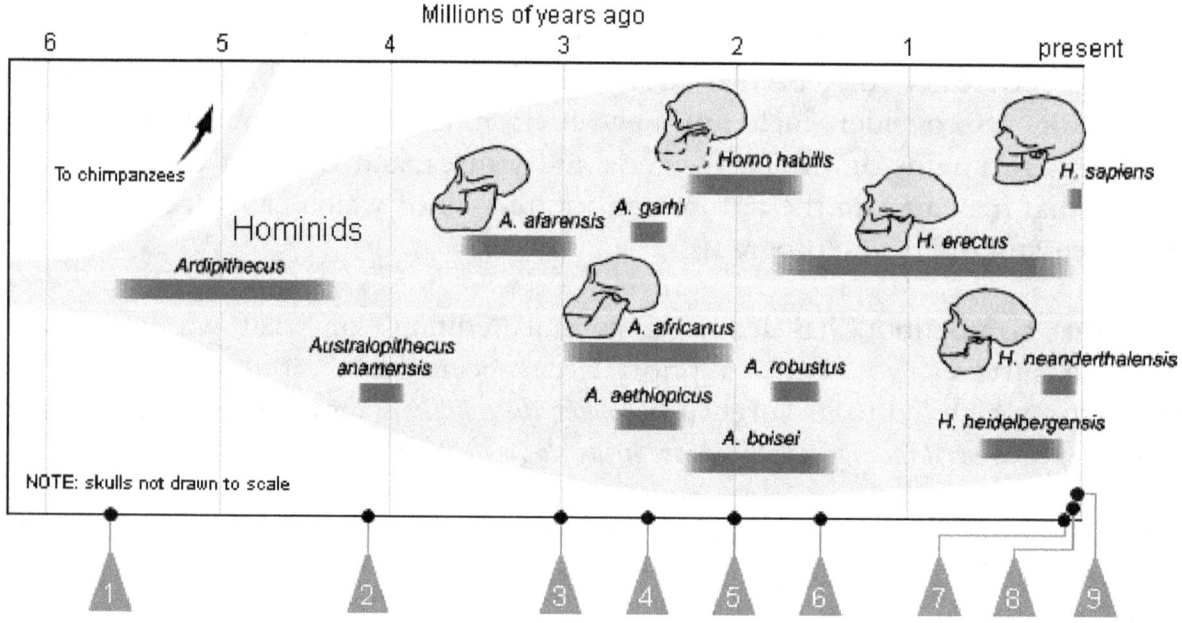

Shaun is literally a f...ing success. Nobody in his genetic past screwed up (pun intended). At each stage where it counted most, Shaun's elemental sequences survived and not once was there a premature extinction or death. If so, he wouldn't be here today. If so, none of us would be alive today. The very fact that you are reading this book can be taken prima facie that you are one of the great success stories in human history, indeed in all of organic history. You have won nature's lottery. . . at least so far.

Imagine how many strands of DNA never made it. Imagine how many countless sperms and eggs never reached fruition. Imagine how many life forms lived and then died before they could transmit their genetic code? Imagine the odds against you and it becomes readily apparent that the very fact you have beat such odds is a clear indication that your unique genome has some fundamental traits that have led to such success. Those traits are called adaptations. Whether you like it or not, you must be sufficiently adapted to your present circumstance or you wouldn't have been able to arrive here.

You are one of the winners.

Given this amazing history and your amazing plasticity to survive nature's twists and turns, the real question that arises is not why but how? What is it about you that has led to you? In other words, you are a distinguished survivor. And how is it that you survived this competition and others did not? Or, more generally, what traits do Homo sapiens possess that has allowed them to last this long?

But before we probe further into that question, another one still begs to be answered. How far back does our genetic code go? Almost all fundamentalist Christians, who believe in a literal interpretation of the Bible, more or less accept evolution by natural selection within a range.

What most of them object to in class is the idea that the human species evolved from some earlier primate species. As one bright Christian argued, "Yes, I can accept micro-evolution, varying adaptations and changes within a species, but what I cannot accept (and which would go against my beliefs) is the idea that one species mutated or evolved into another. I see no evidence of such a thing in the fossil record. Where are all the transitional forms? Human beings are unique."

To tackle this issue head-on in class and to answer the good questions put forth by my student, I first point out what is not so obvious at first. All of evolution, at least in terms of DNA sequencing, is at the micro level.

Instead of thinking of phenotypes (the bodies which house our genetic codes), focus on genotypes which are incredibly small. So small, in fact, that we cannot see even one complicated strand of DNA with our naked eye.

At the molecular level, deoxyribonucleic acid isn't concerned with our macro issues of whether something is a defined species or not. At these tiny scales, it is merely a question of biochemistry. And since humans and chimpanzees and dolphins and bananas all share the same language code (remember, it comes down to a four letter molecular alphabet), at the microscopic level there isn't a thick brick wall dividing DNA into invariant species categorizations which has a sign to all intruding and stray polynucleotides: *Stay Away*. No, rather it is more akin to alphabet cereal (though this cereal only has the letters C, T, A, and G) or alphabet soup where the letters are free to roam wherever they please within the medium of milk or broth. It isn't a question at this realm of a whether a species can mutate into another, but rather if adenine ("A") can bond with thymine ("T") in a complementary base pairing (it can), or if cytosine ("C") can bond in a similar way with guanine ("G")—it can also. In other words, the macro issue of species never emerges at this quartered off biochemical level. Rather, it is a word used after the fact to describe built-up differences of DNA sequencing. The DNA itself is the same and thus the notion of speciation isn't about C or T or A or G, but rather about how these already given molecular clusters form into larger scaffolding projects. Thus if you accept micro-evolution within a species, you have already de facto accepted evolution itself, since all DNA manipulation (which is how evolutionary change occurs over time in organic beings) is at the micro level. As the Brown University *Course on Evolution* explains it:

For evolutionists the revolution in DNA technology has been a major advance. The reason is that the very nature of DNA allows it to be used as a "document" of evolutionary history: comparisons of the DNA sequences of various genes between different organisms can tell us a lot about the relationships of organisms that cannot be correctly inferred from morphology. One definite problem is that the DNA itself is a scattered and fragmentary "document" of history and we have to beware of the effects of changes in the genome that can bias our picture of organismal evolution.

Two general approaches to molecular evolution are to 1) use DNA to study the evolution of organisms (such as population structure, geographic variation and systematics) and to 2) to use different organisms to study the evolution of DNA. To the hard-core molecular evolutionist of the latter type, organisms are just another source of DNA. Our general goal in all this is to infer process from pattern and this applies to the processes of organismal evolution deduced from patterns of DNA variation, and processes of molecular evolution inferred from the patterns of variation in the DNA itself. An important issue is that there are processes of DNA change within the genome that can alter the picture we infer about both organismal and DNA evolution: the genome is fluid and some of the very processes that make genomes "fluid" are of great interest to evolutionary biologists. Thus molecular evolution might be called the "natural history of DNA.

As for why we don't see as many transitional forms as one might expect, this too is a misleadingly framed question, since it implies that such should be easy to find. Quite the opposite is true. The fact that we have found as many as we have is astounding itself, given that the theory of evolution has only been accepted for less than two centuries. And even while accepted in the scientific community, how many researchers are there worldwide trying to unearth these very rare and precious documents of our ancestral past? My husband, David, who is a fond lover of Coca Cola, makes a fitting analogy here. He remembers back in the late 1950s and 1960s when cans of coke did not have the opening tabs we have today. Rather, one had to use a can opener (usually with one opening larger than the other). When can opening tabs were introduced they were clumsy and slightly dangerous. But Coke can tops have undergone an extensive evolution. As one writer on the subject explains it:

In the early 1960's the Pittsburgh Brewing Company introduced "Iron City Beer" in 'self-opening cans.' The concept was pretty novel – just pull up on a tab and you had an open can of beer in your hand! No accessories like a 'church key' or bottle opener necessary - imagine that! These early pull tabs were known as "zip tops" and were disposable. But because of the rough edges of the aluminum, the cans often left people with cuts on their fingers, lips and even noses.

By 1965 the design was changed to the ring style, which I'm sure every metal detectorists has seen his or her share of. The ring style was even easier then the zip top; just put your finger into the ring, yank forward and have your beverage with less potential for physical injury - even better!

Needless to say, the swift evolution of the zip top to the ring tab revolutionized canned beverages. By the mid-60's over 75% of all cans produced in the U.S. had a pull-tab opening.

Ten years after the "ring" version of the pull tab was introduced, an answer to this environmental and safety nightmare finally came. The "stay tab" style was introduced in 1975 by the Falls City Brewing Company, and they were here to stay – literally. These ring-style-stay-tabs are what we can see on every can of coke and beer in the grocery store today. Unfortunately, they don't stay quite as well as the designers would have liked. But at least this style doesn't force people to throw the tab aside… they actually have to do a little work to get it off.

However, today there are many students in college who are unaware of the evolution behind opening tabs on coke cans and other soft drinks. Indeed, there haven't been just three stages in this evolution but many small incremental changes, most of which have gone unnoticed. If you look just at a Coke can of the 1950s and a Coke can of today, you might ask where did all the transitional forms go? How easy would it be to find each and every modification over the last fifty or so years?

Unless you are an avid collector it wouldn't be easy at all, since most of the cans have been discarded and thus their respective histories have been smashed or buried. I bring up this analogy because there are millions of such canned fossils waiting to be found but it would take inordinate amounts of time and patience to unearth them, if one hadn't already kept a record of it as the cans were improved over time--and this is about an object for which we have tremendous amounts of information. Imagine how difficult it must be to find transitional forms of our ancestors that lived millions of years ago? So many conditions have to be right for us to be lucky enough to find even one example, much less several.

Kathleen Hunt elaborates on this her website on transitional forms in the fossil record:

The first and most major reason for gaps is "stratigraphic discontinuities", meaning that fossil-bearing strata are not at all continuous. There are often large time breaks from one stratum to the next, and there are even some times for which no fossil strata have been found. For instance, the Aalenian (mid-Jurassic) has shown no known tetrapod fossils anywhere in the world, and other

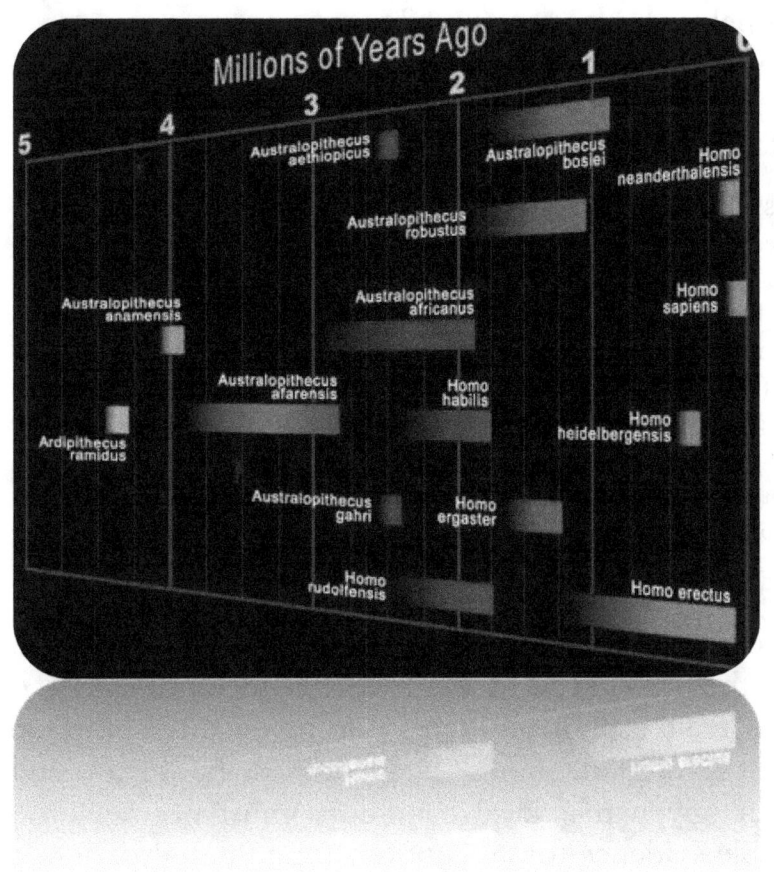

stratigraphic stages in the Carboniferous, Jurassic, and Cretaceous have produced only a few mangled tetrapods. Most other strata have produced at least one fossil from between 50% and 100% of the vertebrate families that we know had already arisen by then (Benton, 1989) -- so the vertebrate record at the family level is only about 75% complete, and much less complete at the genus or species level. (One study estimated that we may have fossils from as little as 3% of the species that existed in the Eocene!) This, obviously, is the major reason for a break in a general lineage. To further complicate the picture, certain types of animals tend not to get fossilized -- terrestrial animals, small animals, fragile animals, and forest-dwellers are worst. And finally, fossils from very early times just don't survive the passage of eons very well, what with all the folding, crushing, and melting that goes on. Due to these facts of life and death, there will always be some major breaks in the fossil record. Species-to-species transitions are even harder to document. To demonstrate anything about how a species arose, whether it arose gradually or suddenly, you need exceptionally complete strata, with many dead animals buried under constant, rapid sedimentation. This is rare for terrestrial animals. Even the famous Clark's Fork (Wyoming) site, known for its fine Eocene mammal transitions, only has about one fossil per lineage about every 27,000 years. Luckily, this is enough to record most episodes of evolutionary change (provided that they occurred at Clark's Fork Basin and not somewhere else), though it misses the most rapid evolutionary bursts. In general, in order to document transitions between species, you specimens separated by only tens of thousands of years (e.g. every 20,000-80,000 years). If you have only one specimen for hundreds of thousands of years (e.g. every 500,000 years), you can usually determine the order of species, but not the transitions between species. If you have a specimen every million years, you can get the order of genera, but not which species were involved. And so on. These are rough estimates (from Gingerich, 1976, 1980) but should give an idea of the completeness required. Note that fossils separated by more than about a hundred thousand years cannot show anything about how a species arose….

But even with these severe limitations, archaeologists have already unearthed a number of very impressive transitional fossil remains. This is quite remarkable, as we have pointed out, given the inordinate difficulty inherent in trying to discover biological remnants that are still intact.

Darwin's DNA

Here is just a partial list of transitional forms among amphibians:

Temnospondyls, e.g Pholidogaster (Mississippian, about 330 Ma) -- A group of large labrinthodont amphibians, transitional between the early amphibians (the ichthyostegids, described above) and later amphibians such as rhachitomes and anthracosaurs. Probably also gave rise to modern amphibians (the Lissamphibia) via this chain of six temnospondyl genera, showing progressive modification of the palate, dentition, ear, and pectoral girdle, with steady reduction in body size (Milner, in Benton 1988). Notice, though, that the times are out of order, though they are all from the Pennsylvanian and early Permian. Either some of the "Permian" genera arose earlier, in the Pennsylvanian (quite likely), and/or some of these genera are "cousins", not direct ancestors (also quite likely). Dendrerpeton acadianum (early Penn.) -- 4-toed hand, ribs straight, etc. Archegosaurus decheni (early Permian) -- Intertemporals lost, etc. Eryops megacephalus (late Penn.) -- Occipital condyle splitting in 2, etc. Trematops spp. (late Permian) -- Eardrum like modern amphibians, etc. Amphibamus lyelli (mid-Penn.) -- Double occipital condyles, ribs very small, etc. Doleserpeton annectens or perhaps Schoenfelderpeton (both early Permian) -- First pedicellate teeth! (a classic trait of modern amphibians), etc.

We have a mistaken notion about evolution because we tend to think only at the level of large body types, forgetting that the real changes occur at the biochemical level and even the smallest change there can have a dramatic impact on its eventual housing. While morphological evidences of evolution should by definition be scarce and difficult to precisely piece together (given that ideal conditions must be met on a series of fronts), the most remarkable evidence for evolution is found exactly where it should be uncovered: at the level of DNA.

Sean Carroll has written a popular account of this wonderful breakthrough in evolutionary biology entitled *The Making of the Fittest*. The Howard Hughes Medical Institute provides a nice summary of Carroll's work:

For decades, scientists studying evolution have relied on fossil records and animal morphology to painstakingly piece together the puzzle of how animals evolved. Today, growing numbers of scientists are using DNA evidence collected from modern animals to look back hundreds of millions of years to a time when animals first began to evolve. One of those leading the charge is molecular biologist Sean Carroll.

Carroll's research focuses on the way new animal forms have evolved, and his studies of a wide variety of animal species have dramatically changed the face of evolutionary biology. Using genetics and the tools of molecular biology, he is looking back to the dawn of animal life some 600 to 700 million years ago. It is so long ago that there are virtually no fossils or other physical clues to indicate what Earth's earliest animals were like.

Darwin's DNA

"Evolution encompasses all of biology—it is our big picture," Carroll said. "When I was a student, we had a grand picture of animal evolution from the fossil record, but no knowledge whatsoever of how new animal forms arose. That is the mystery that I want to tackle."

Carroll's studies have uncovered evidence that an ancient common ancestor—a worm-like animal from which most of the world's animals evolved—had a set of "master" genes to grow appendages, such as legs, arms, claws, fins, and antennas. Moreover, Carroll noted, these genes were operational at least 600 million years ago and are similar in all animals, from humans to vertebrates, insects, and fish. What is different, however, is the way these genes are expressed, leading some animals to develop wings, and others to grow claws or feet.

"We found the same mechanism in all the divisions of the animal kingdom," Carroll noted. "The architecture varies tremendously, but the genetic instructions are the same and have been preserved for a very long period of time."

Carroll is also probing the common fruit fly, Drosophila melanogaster, to elucidate how genes control the development and evolution of animal morphology, or form. This innovative approach to studying evolution has led scientists to a more detailed understanding of how animal patterns and diversity evolve.

By analyzing the genetic origin of the decorative spots on a fruit fly wing, Carroll has discovered a molecular mechanism that helps to explain how new patterns emerge. The key appears to lie in specific segments of DNA, rather than genes themselves, that dictate when during development and where on an insect's body proteins are produced to create spots or other patterns.

The same molecular mechanism is likely at work in other animals, including humans, and helps to explain the pattern of stripes on a zebra or the technicolor tail of the peacock. Carroll and his colleagues chose to study the evolution of the wing spot on fruit flies because it is a simple trait with a well-understood evolutionary history. While ancient fruit fly species lack spots, some species have evolved spots under the pressure of sexual selection. The wing spots offer a survival advantage to males, who depend on the decorations to "impress" females to choose them in the mating process.

The discovery is important because it provides critical evidence of the way that animals evolve new features to improve their chances of reproductive success and survival. "We now have convincing proof that evolution occurs when accidental mutations create features such as spots or stripes that impart an advantage for attracting mates, hiding from or confusing predators, or gaining access to food," Carroll explained. "These accidents are then preserved as small changes in the DNA."

At this point in the lecture, most of my students are nodding their head about the logic of evolution, even though they may not agree with all the pointed details.

Darwin's DNA

Evolution by natural selection is, as Daniel Dennett rightly pointed out in his book *Darwin's Dangerous Idea,* based on the notion of methodological naturalism, whereby one attempts to explain all phenomena by its constituent parts. Paul and Patricia Churchland have called this approach intertheoretic reductionism. Take any physical object and you have two fundamental options in trying to explain it. Either it is material or it is not. If the former you try to ground your explanations in physics, chemistry, biology, psychology, and sociology—with an eye and ear to the ground from which these emergent structures arise. If the latter, you are engaged in a metaphysical enterprise, where things are explained not by other material substances but transcendent, even spiritual, realities. Dennett has invoked a nice metaphor to explain these different approaches: science is a crane like approach, and follows an algorithmic (step by step procedure) mindset, even if one is allowed all sorts of wild imaginings provided they are ultimately tested and verified by empirical experiments. Religion, on the other hand, is a sky hook and tends toward non-algorithmic explanations.

This is why evolution is such a powerful idea. It explains so much so simply. Dennett has called it the single greatest idea in the history of human thought, since it serves as the backbone for almost every one of the sciences—from astronomy to neuroscience. As Theodosius Dobzhansky, one of the architects of the neo-synthesis of evolutionary theory in the mid-20th century points out, "Nothing in biology makes sense except in the light of evolution."

It is right at this juncture that I raise a larger philosophical issue in my lecture. If evolution by natural selection (and other selective or eliminative forces) can indeed explain why Shaun looks the way he does, can it also help explain why Shaun thinks the way he does?

Explaining Consciousness

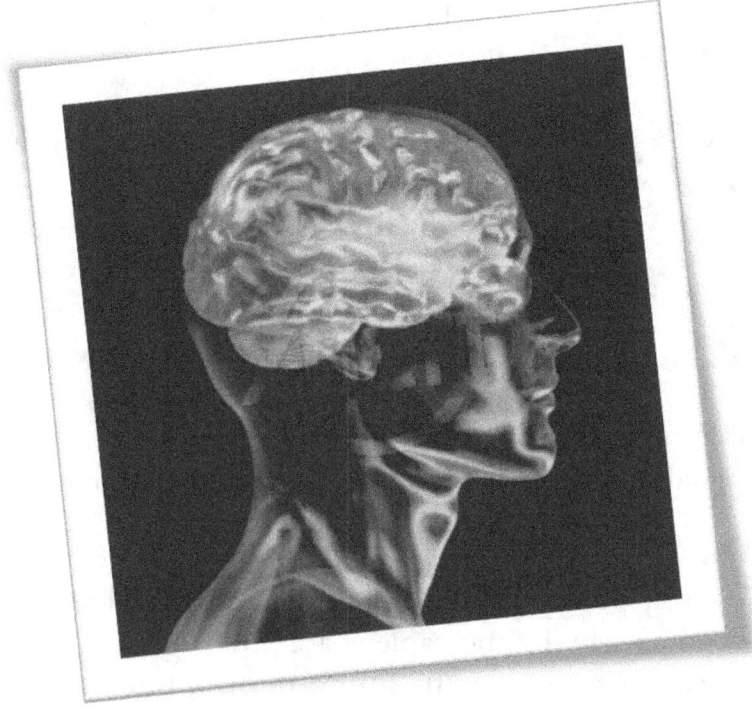

In the distant future I see open fields for far more important researches. Psychology will be based on a new foundation, that of the necessary acquirement of each mental power and capacity by gradation. Light will be thrown on the origin of man and his history.
—Charles Darwin

Why do we think the way we do? Or, more in line with the theme of philosophy, why does the question "why" arise so much in human beings?

There have been, to be sure, many answers to these queries from time immemorial. Countless religions have been created to resolve these perennial questions, each with differing successes. Even science has gotten into the fray with the blossoming of psychology as a distinct discipline in the latter part of the 19th century.

Charles Darwin made a very pregnant prediction near the end of his classic 1859 tome when he opined that evolutionary theory would radically transform other fields of research. He even went so far as to say that "Psychology will be based on a new foundation, that of the necessary acquirement of each mental power and gradation." He could have just as easily substituted Philosophy.

While there have been some remarkable developments in evolutionary psychology, a field previously known more controversially as sociobiology, there hasn't been the same attention given to philosophy. Historically, this may be due to the fact that Herbert Spencer, an early champion of fusing philosophy and evolution and a quite popular advocate of such during his lifetime, became something of anathema during the latter part of the 19th and early 20th century because of some of his more controversial views, particularly Social Darwinism. As the entry on him in *Wikipedia* notes: "Posterity has not been kind to Spencer. Soon after his death his philosophical reputation went into a sharp and irrevocable decline. Half a century after his death his work was dismissed as a 'parody of philosophy' and the historian Richard Hofstadter

called him the 'the metaphysician of the homemade intellectual and the prophet of the cracker-barrel agnostic.'"

Combining philosophy with evolution can be fraught with peculiar dangers, not the least of which is a tendency towards what Dennett has called "cheap reductionism," explaining away complex phenomena instead of properly understanding it. Nevertheless, it is even more troublesome to ignore Darwinian evolution because it illuminates so many hitherto intractable problems ranging from epistemology to ethics.

The new field of evolutionary philosophy, unlike its aborted predecessors of the past, is primarily concerned with understanding why Homo sapiens are philosophical in the first place. It is not focused on advocating some specific future reform, but rather in uncovering why humans are predisposed to ask so many questions which, at least at the present stage, cannot be answered. In other words, if evolution is about living long enough to transmit one's genetic code, how does philosophy help in our global struggle for existence?

To answer that question and others branched with it, one has to deal with the most complex physical structure in the universe—the human brain. Because it is from this wonder tissue, what Patricia Churchland has aptly called "three pounds of glorious meat," that all of human thought, including our deep and ponderous musings, is built upon. Take away the human brain and you take away all of philosophy.

Therefore, if we are to understand why philosophy arose in the first place, we have to begin with delving into the mystery on why consciousness itself arose. And to answer that question we first have to come to grips with Darwin's major contribution to evolutionary theory—natural selection. Why would nature select for awareness, especially the kind of self-conscious awareness endemic to human beings, when survival for almost all species is predicated upon unconscious instincts? What kind of advantage does self-reflective consciousness confer that would allow it to emerge and develop over time?

I thought long and hard over this question and for many years I never found a satisfactory response. It was only when I began working as a Research Assistant for Professor V.S. Ramachandran at UCSD, where we focused on visual perception, that I realized that the answer to consciousness was intimately related to Darwinian natural selection. When I started to think in evolutionary terms, it became clearer to me that I had been asking the wrong sorts of questions, especially as I tended to relate my philosophical musings with my religious upbringing.

It may be simply a coincidence but I noticed that after I had met with Francis Crick at a dinner party hosted by Ramachandran in his La Jolla home my questions became more focused and thus I proceeded upon a more fruitful line of inquiry. Ironically, I didn't know that the person I was introduced to at the party was a world famous scientist who along with James Watson had discovered the double-helix structure of DNA. Ramachandran simply introduced the Nobel laureate as Francis. Since we were the first guests at the party we ended up chatting on the nearby couch. However, we both became so absorbed in our conversation that we ended up talking for nearly two straight hours. At the time I thought Francis was a second-fiddle to his artist wife, since he never gave any indication of his remarkable credentials. For all I knew he was either retired or out of a job. When I asked Crick what he did for a living he smirked and said, "Ah, I dabble in a little bit of this and a little bit of that," never once revealing his vast background in molecular biology.

Francis learned of my interest in Eastern Philosophy, Gnosticism, and my advocacy of vegetarianism. He tried to provoke me a bit, but because I didn't think he was all that knowledgeable in these areas I passionately, but hopefully reasonably, gave him my counter-arguments. Crick seemed slightly bemused by my candor and complete lack of awe in his presence. I even had the audacity of suggesting that he could help me out in the lab at UCSD with the experiments we were doing. It was only later to my and everyone else's chagrin when I realized (and to my horror) that I was debating

with such an eminent mind. Thankfully, and to his great credit, Francis Crick never let on during the entire conversation who he was.

What stood out to me both during and after our conversation was Crick's insistence that I should focus my graduate studies on neuroscience, since he argued that the brain was the key to unraveling many of life's greatest mysteries. Although I didn't at the time take Francis' advice, his suggestion stayed with me over the years and has clearly influenced my thinking on how to approach the study of consciousness.

My husband (who did his Ph.D. at UCSD) was also interested in why consciousness evolved and had since 1991 embarked on an intensive study of physics and neuroscience to better resolve the issue.

I still remember the day he came home from his teaching duties at CSULB and exclaimed that the answer to the riddle of consciousness was remarkably simple and obvious. So obvious, in fact, that he wondered why he hadn't dawned on him much earlier. In his typically Socratic way, he peppered me with a series of loaded questions designed to make clearer his epiphany. Watching our son growing up, said my husband, was the triggering event. As my husband recounted in his diary of that day,

I am learning more about the human brain and philosophy from Shaun than I ever did from books. It is now very obvious and clear to me that whatever questions we ask of the universe arise because of the architecture of our brain. More precisely, philosophy is the result of differing brain states and upon that contingent scaffolding we come up with varying questions to ask of the world and its participants, though we never seem to realize that such questioning has less to do with reality per se and more to do with our own evolutionary needs. Ah, I can put it better yet: philosophy is like heartburn. It is the natural result of something that didn't digest well. I will call it brain burn.

What does such a neologism mean? Every deep question we have, every deep thought we ponder, is the result of the confusion of a neural system when confronted with its own dissociation. Consciousness is dissociation. And therein lays its Darwinian advantage, since most of our awareness is in our head it doesn't have to face the very real and empirical and deathly consequences of being without. Being within survives. Being without tends to end up dead. So consciousness

| Darwin's DNA

arises as dissociation so it can play out (via its internal machinations. . . what we call imagination/daydreaming) without physical harm alternative scenarios to secure its Four F'S: *F..k, Food, Flee, Fight*. Consciousness is literally a virtual simulator and that is why it has been so helpful in allowing humans to survive globally, even when our bodies were not adapted to certain environmental niches.

If you can imagine without real consequence, then you have a better chance of living if you have already played out (internally, but not externally) competing strategies. Those without consciousness don't have this liberty and thus when they do play out a choice, they so in a real world. And in such a real world, if it doesn't work you are eaten. In imagination, in consciousness, you can play as if it is real and project all sorts of end game earnings to see which one would be to your advantage. Consciousness is the brain's way of making chance/chaos (read nature) more plastic, more pliable, more beneficial to the host organism.

What is the best way to survive chance contingencies? By developing a statistically deep understanding of what varying options portend. Consciousness is a way around pure chance by developing an internalized map of probabilities which can be visualized internally without having to be outsourced prematurely. Being in your head is another way of avoiding being stuck in only a one way avenue of recourse. Any reproducing DNA that can develop a virtual simulator within itself has a huge advantage over a genetic strand that cannot.

The gurus and mystics have it completely wrong. The world isn't an illusion; consciousness is. But even if we say consciousness is an illusion--which it is (in the sense that its stuff is literally composed of dreams and other sorts puffery)--that doesn't mean it isn't helpful to our survival. It is. Consciousness is a body's way of giving itself a better chance when confronted with the reality of Chance itself. Consciousness is probability functions envisioned.

Ah, but we humans are so naive. We were so mesmerized by the theatre of consciousness that we forgot that it was the body that was real, not its projected sideshow. So I won't be misunderstood. What I am saying is the direct opposite of religion, of mysticism, of guruism. Consciousness isn't the strongest part of a human being. It is not going to survive. There is no soul. Nietzsche was right. Consciousness is the weakest element of a human being, though we believe otherwise. Film is fragile, so is our awareness. Nobody believes that a reel of film is going to reincarnate or survive in the afterlife. Awareness is Poker Self-Reflective. Chance giving itself better odds.

Darwin's DNA

My husband had been reading various articles by Ramachandran and a number of books by the Nobel laureate in medicine, Gerald Edelman (most popularly known for his introduction of Neural Darwinism) where they had implied that consciousness was more or less a virtual simulator. "That's it," Dave echoed. "Think about what consciousness does most of the time. It simulates its environment in order to make the best probable guesses about how to respond to it. But in order for such simulations to be wide-ranging, there has to be an allowance for lots of projections or end game results which are purely fantastical and have no concordance with empirical reality. Otherwise, the very basis of imagination would be too stilted and wouldn't be amenable to new situations, new environments, and new problems."

At this point I chimed in with Stephen Jay Gould's notion of spandrels, the unintended consequences or secondary effects of a more primary adaptation. If consciousness is an evolutionary adaptation which allows for any sophisticated strand of DNA to develop a virtual navigating device within itself, thereby increasing its odds by allowing for a prior contemplation of varying strategies before making a decisive and decidedly empirical decision, then it is quite reasonable to expect that there should be much in a virtual simulator which is merely imaginary and has no real world correlate. This would easily explain why many of our projections are delusional. If consciousness is a probability adjustor (played out in our minds replete with an emotional feedback loop), then sometimes we will opt for strategies that are indeed wrong, misguided, and illusionary. They are just part of the odds. If this is true, then we should expect certain mental states which would deviate from the mean. Is this the basis for all mental diseases?

But as I was partnering these ideas with my husband, I realized that even calling something a mental disease implied an already fixed understanding of what constituted normal awareness. If consciousness is simulating its environment so as to better its odds when it does indeed make a real life choice, then much of its success is due to how well it actually matches up with and predicts the incoming stimuli. Consciousness could never have survived the brutal machinations of natural selection unless it was somewhat accurate in how it modeled its exterior environment. If it came up with simulations that were continually mistaken, it would have been eaten up

by a predator long ago. No, consciousness must be on whole a fairly accurate modeler of the outside world in order to confer the benefit necessary to evolve as a significant adaptation. That there will be glitches and mistakes and illusory detours is to be expected, but overall those have to be kept at a minimum since otherwise the very advantage that consciousness confers as a virtual simulator would be automatically lost. If consciousness was merely projections of our temporal lobes, it wouldn't have any real world benefit, but would act as a punishing (and ultimately eliminating) detriment.

As we discussed this theory of consciousness further, we started to see the powerful utility of Gerald Edelman's binary idea of *Second Nature*. Edelman makes a distinction between first order and second order consciousness, or what he calls first nature and second nature levels of awareness. As the *Wilson Quarterly* review of his book, *Second Nature*, explains:

In *Second Nature*, Nobel Prize winning neuroscientist Gerald Edelman proposes what he calls "brain-based epistemology," which aims at solving the mystery of how we acquire knowledge by grounding it in an understanding of how the brain works.

Edelman's title is, in part, meant "to call attention to the fact that our thoughts often float free of our realistic descriptions of nature," even as he sets out to explore how the mind and the body interact.

Edelman suggests that thanks to the recent development of instruments capable of measuring brain structure within millimeters and brain activity within milliseconds, perceptions, thoughts, memories, willed acts, and other mind matters traditionally considered private and impenetrable to scientific scrutiny now can be correlated with brain activity. Our consciousness (a "first person affair" displaying intentionality, reflecting beliefs and desires, etc.), our creativity, even our value systems, have a basis in brain function.

The author describes three unifying insights that correlate mind matters with brain activity. First, even distant neurons will establish meaningful connections (circuits) if their firing patterns are synchronized: "Neurons that fire together wire together." Second, experience can either strengthen or weaken synapses (neuronal connections). Edelman uses the analogy of a police officer stationed at a synapse who either facilitates or reduces the traffic from one neuron to another. The result of these first two phenomena is that some neural circuits end up being favored over others.

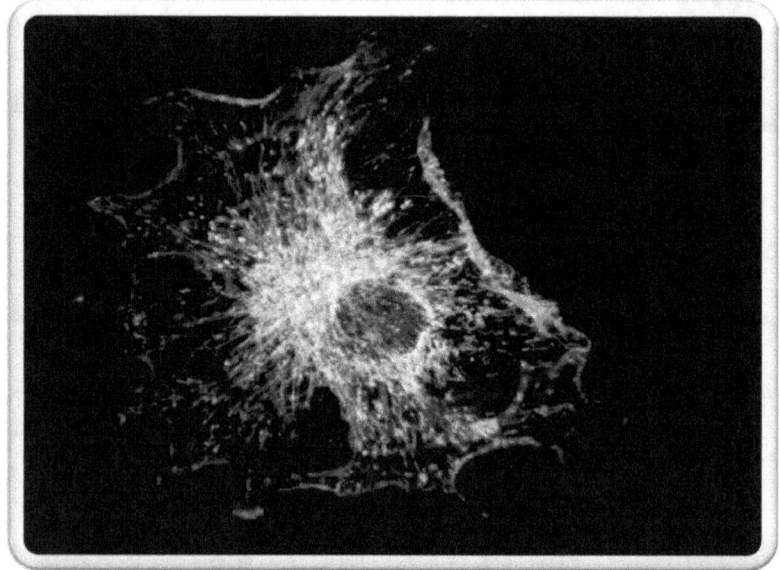

Finally, there is reentry, the continued signaling from one brain region to another and back again along massively parallel nerve fibers. Since reentry isn't an easy concept to grasp, Edelman again resorts to analogy, with particular adeptness: "Consider a hypothetical string quartet made up of willful musicians. Each plays his or her own tune with different rhythm. Now connect the bodies of all the players with very fine threads (many of them to all body parts). As each player moves, he or she will unconsciously send waves of movement to the others. In a short time, the rhythm and to some extent the melodies will become more coherent. The dynamics will continue, leading to new coherent output. Something like this also occurs in jazz improvisation, of course without the threads!" Reentry allows for distant nerve cells to influence one another: "Memory, imaging, and thought itself all depend on the brain 'speaking to itself.'"

In Edelman's view sentient creatures have (to greater or lesser degrees) first order awareness or consciousness, including human beings. But such awareness is merely rudimentary and is not yet self reflective; it is more akin to being aware of something exterior to one's self, but not yet being able to reflect upon what that means. It is like looking at a mirror but without any comprehension of what that reflection is or means. Second nature, or second order awareness, is when we look at a mirror and develop a feedback loop and have the ability to ponder over what it means. If first nature is a way for our senses to reach out to the world and receive stimuli and have our instincts respond accordingly, second nature is our ability to absorb such information and have the wherewithal to reconstruct models of varying probabilities about what this information means. Second nature allows us to simulate our environment in ways not possible with first order awareness. In this sense, it confers a dramatic Darwinian advantage because simulations allow for better odds in our ultimate reactions to whatever stimuli or information we encounter.

As the Edelman entry on *Answers.com* explains further:

Edelman was struck by a number of similarities between the immune system and the nervous system. Just as a lymphocyte can recognize and respond to a new antigen, the nervous system can respond similarly to novel stimuli. Neural mechanisms are selected, he argued, in the same manner as antibodies. Although the 109 cells of the nervous system do not replicate, there is considerable scope for development and variation in the connections that form between the cells. Frequently used connections will be selected, others will decay or be diverted to other uses. There are two kinds of selection: developmental, which takes place before birth, and experiential. There are also innate 'values' – built in preferences for such features as light and warmth over the dark and the cold.

In Edelman's model, higher consciousness, including self-awareness and the ability to create scenes in the mind, have required the emergence during evolution of a new neuronal circuit. To remember a chair or one's grandmother is not to recall a bit of coded data from a specific location; it is rather to create a unity out of scattered mappings, a process called by Edelman a 'reentry'. Edelman's views have been dismissed by many as obscure; some neurologists, however, consider Edelman to have begun what will eventually turn out to be a major revolution in the neurosciences.

As the Guardian newspaper elaborates:

The degree to which an organism is conscious is therefore dependent on the complexity of its brain. Large-brained mammals such as dogs have a core self-consciousness. Humans, perhaps uniquely, have a reflexive and recursive consciousness - we are conscious of being conscious.

Human consciousness is thus an evolved property, the inevitable consequence of having brains of a particular complexity.

What David and I found most intriguing in Edelman's theory was his fundamental understanding what of consciousness does. As Steven Rose summarizes Edelman's definition, "What consciousness does, he says, is to "inform us of our brain states and is thus central to our understanding."

Awareness, therefore, is indeed a theatre for the mind in which varying trajectories are given stage time to play out their respective hunches. But because this is a theatre and not the outside world as such, many things are acted out as if they were indeed real. And herein lays the ultimate danger of such an awareness. One can too easily conflate one's neural state for the real state of the world at large. Freud, whom Edelman both praises and criticizes, often remarked that one of his greatest discoveries was psychic transference. As Alun Jones explains:

Freud described transference as "new editions and facsimiles of impulses and phantasies" (1923/1953, p. 82) originating in the past. Instead of remembering, the person transfer attitudes and conflicts are enacted in current relationships, sometimes with unfortunate results. Manifestations are likely to occur in all human encounters; feelings toward the significant other often begin to emerge early on in relationships.

Freud's notion of transference is what consciousness was evolved to do. For instance, in our early childhood our awareness developed several models about what was happening, especially in a situational crisis. If that model was believed and sustained to a given extent, then it is quite likely to be invoked again if a similar occasion arose. The problem, of course, is that the new dilemma may have little resemblance to the prior one and thus our projected transference may be of little help. But even here the transference is of use psychologically if we can see it for it was and is: "information about our brain state and its attendant emotions."

While Edelman rightly discounts Freud's psychoanalytic technique, he does nevertheless appreciate Freud's rich and nuanced understanding of how consciousness may operate. Freud may have been technically incorrect in his analysis of why such mental problems arise and how to resolve them, but he was an insightful cartographer of psychic ailments.

The problem of consciousness is, as Freud rightly suspected, revealed by dreaming, since it gives us a clue about how consciousness must have first emerged. As we all know there are several stages of awareness in our dreams—ranging from the hallucinatory and intangible impressions one might get from a severe lack of rest to the elevated heights of lucidity when one experiences an acute sense of clarity and self luminosity. Our waking state awareness is built upon the feeling of certainty, and to the degree that we can believe what our mind projects the more likely we are to act upon it, rightly or wrongly. Thus the hallmark of a heightened sense of awareness is how certain and definite it feels. This is an important feature to differentiate since if we lacked this clarity we would be less willing to act and to make choices. And if we hesitate in face of a real danger (a lion, a tiger, a bear, oh my!), then we run the very real risk of being eaten. An awareness that couldn't make such fine tuned adjustments would have been eliminated long ago in our ancestral past.

But herein lies the problem. To the degree that consciousness "works" because it conflates our brain state with the current "real" state of the world around us, it runs the risk of not being pliable enough to adapt to a new set of circumstances. In other words, consciousness must have enough "plasticity" to make judgments that turn out to be wrong, provided those mistaken assumptions error on the side of being too conservative or too safe.

A classic example of this, albeit wildly over stated here for our purposes, is the hypothetical case of an animator versus a non-animator. Imagine for argument's sake that hundreds of thousands of years ago there existed two kinds of human beings. One who tended to believe in animism, whereby ordinary objects possessed an "animated" or "vital" or "soul" force replete with the same kind of intentionality, though perhaps more mysterious, than we have. The other human being lacked such animating tendencies and thus had a more limited way of deciphering the motivations or intentions of other objects. Now, let's further imagine that both of these individuals were brothers sitting around a camp fire and there was the sound of a rustling bush. The animator, given his predisposition to project (Freud's transference writ early?), imagines all sorts of possibilities and because of the emotions he attaches to each (and keeping in mind the conservative nature of how such a state of awareness must have in order to survive) will on average try to protect himself and thus run and hide or protect himself under the worst case scenario. The non-animator, however, will do no such thing and in his "realist" approach most likely stay put and not animate anything whatsoever about the rustling bushes.

Now it may well be that the non-animator is correct much more often than the animator (most of the time it is just the wind against the leaves), but if the animator is right just 1/10 of the time his conservative, though animistic, approach has saved his life and thereby allowed him to live another day in order to pass on his genetic code. The non-animator, even if he was correct more often than his counterpart, ends up eaten and dead, and thus reducing his chances to pass on his DNA.

While this example is hyperbolic to the extreme, it does underline something about animism and about why such an innate tendency may have evolved in human beings. Projections are future oriented and any awareness that can better predict the future will be of some survival benefit. But the future is unknown, and therefore making predictions about it is inherently a probabilistic venture. Therefore if consciousness arose at least to some small measure as a predictor of how to react to future but unknown events, then it must error on the side of caution. Otherwise, if awareness was too adventurous whatever advantages it gained would be prematurely lost with one bad outcome.

But this conservative approach of awareness would most likely only be necessary when an individual confronted what he or she perceived to be a "real" or "certain" crisis. In other words, a virtual simulator would only be cautious if it acted upon that simulation as if it related to a real event. If, however, the virtual simulator knew or believed beforehand that the simulation wasn't going to be invoked immediately and thus could be contemplated upon without any empirical test, then it could allow itself the freedom of a wide arrange of imaginings. It could, to invoke a cliché, "day dream." It could fantasize. It could conjure up all sorts of nonsense, provided that it wasn't forced into an early test case.

Here we are starting to see what dreaming portended for waking state awareness. Dreams, by definition, are subjective conjurings that arise when we are asleep. This is another way of saying that when consciousness doesn't have to "work" or be called into the line of fire, it has the freedom to mix and match all sorts of images and sounds and feelings into a Picasso like universe.

This would be the precursor of a consciousness which had to act upon one of its modeling simulations. Dream first, act later. And this is apparently what consciousness has evolved to do and why it has been of such a huge advantage to Homo sapiens. Unlike other animals, which apparently do not possess Edleman's second nature, human beings can play out innumerable scenarios in the privacy and safety of their own head until such time that they can draw upon this rich rolodex of imagined trajectories and select what he or she believes is the best approach or model to apply to the present circumstance or problem.

To appreciate how effective and powerful this tool can be for any competitive organism, just ask yourself this question: Who would your rather have fly your airplane: a pilot who never underwent any flight simulations or one who underwent hundreds of hours in a flight simulator? The answer is obvious. Indeed, one can draw from many other professions to see the inherent advantages to virtual modeling or simulation. Most of the top athletes in the world today—from Tiger Woods to Kelly Slater to Michael Jordan—have repeatedly emphasized the importance of visualizing their performance before it happens. Whether it is going over and over in your head how to shoot a basketball in a hoop, or how to stand up quickly to ride a wave at Pipeline, or how to line up a long putt, the more time that is spent simulating the event the better off one feels when actually confronting the real occasion.

So if consciousness did indeed evolve over long epochs of time, we should expect to see varying gradations of awareness depending more or less on the complexity of the neuronal architecture that gives rise to it. Human awareness, though appearing unique and distinct, represents one end of the spectrum of consciousness. If awareness, as Crick and others points out, is intimately connected how our brain functions, then we should expect to see forms of higher awareness or even cognitive function in mammals closely aligned with us from our evolutionary past. And as recent studies in animal behavior have shown, this is precisely what researchers have found. As Donald R. Griffin points out in *Animal Minds*:

Must we reject, or repress, any suggestion that the chimpanzees or the herons think consciously about the tasty food they manage to obtain by these coordinated actions? Many animals adapt their behavior to the challenges they face either under natural conditions or in laboratory experiments. This has persuaded many scientists that some sort of cognition must be required to orchestrate such versatile behavior. For example, in other parts of Africa chimpanzees select suitable branches from which they break off twigs to produce a slender probe, which they carry some distance to poke it into a termite nest and eat the termites clinging to it as it is withdrawn. Apes have also learned to use artificial communication systems to ask for objects and activities they want and to answer simple questions about pictures of familiar things. Vervet monkeys employ different alarm calls to inform their companions about particular types of predator.

Such ingenuity is not limited to primates. Lionesses sometimes cooperate in surrounding prey or drive prey toward a companion waiting in a concealed position. Captive beaver have modified their customary patterns of lodge- and dam-building behavior by piling material around a vertical pole at the top of which was located food that they could not otherwise reach. They are also very ingenious at plugging water leaks, sometimes cutting pieces of wood to fit a particular hole through which water is escaping. Under natural conditions, in late winter some beaver cut holes in the dams they have previously constructed, causing the water level to drop, which allows them to swim about under the ice without holding their breath.

Nor is appropriate adaptation of complex behavior to changing circumstances a mammalian monopoly. Bowerbirds construct and decorate bowers that help them attract females for mating. Plovers carry out injury-simulating distraction displays that lead predators away from their eggs or young, and they adjust these displays according to the intruder's behavior. A parrot uses imitations of spoken English words to ask for things he wants to play with and to answer simple questions such as whether two objects are the same or different, or whether they differ in shape or color. Even certain insects, specifically the honeybees, employ symbolic gestures to communicate the direction and distance their sisters must fly to reach food or other things that are important to the colony.

Although these are not routine everyday occurrences, the fact that animals are capable of such versatility has led to a subtle shift on the part of some scientists concerned with animal behavior. Rather than insisting that animals do not think at all, many scientists now believe that they sometimes experience at least simple thoughts, although these thoughts are probably different from any of ours. For example, Terrace (1987, 135) closed a discussion of "thoughts without words" as follows: "Now that there are strong grounds to dispute Descartes' contention that animals lack the ability to think, we have to ask just how animals do think." Because so many cognitive processes are now believed to occur in animal brains, it is more and more difficult to cling to the conviction that this cognition is never accompanied by conscious thoughts.

Conscious thinking may well be a core function of central nervous systems. For conscious animals enjoy the advantage of being able to think about alternative actions and select behavior they believe will get them what they want or help them avoid what they dislike or fear. Of course, human consciousness is astronomically more complex and versatile than any conceivable animal thinking, but the basic question addressed in this book is whether the difference is qualitative and absolute or whether animals are conscious even though the content of their consciousness is undoubtedly limited and very likely quite different from ours. There is of course no reason to suppose that any animal is always conscious of everything it is doing, for we are entirely unaware of many complex activities of our bodies. Consciousness may occur only rarely in some species and not at all in others, and even animals that are sometimes aware of events that are important in their lives may be incapable of understanding many other facts and relationships. But the capability of conscious awareness under some conditions may well be so essential that it is the sine qua non of animal life, even for the smallest and simplest animals that have any central nervous system at all. When the whole system is small, this core function may therefore be a larger fraction of the whole.

Now turning our attention directly to philosophy we are in a better position to understand why the question "why" arises so often in human beings. In light of consciousness as a virtual simulator, any organism that can develop a mental "pivot" tool will have a tremendous advantage in thinking of new and unexpected strategies.

A curious, but hopefully, useful analogy can be derived here from a well known sport. In basketball, for instance, a seasoned player knows well how to use his or her pivot "foot." Once one has finished dribbling the ball, he or she must keep one foot firmly set on the ground. The other foot, however, is free to "pivot" or "revolve" or "turn" giving one options that the other foot doesn't.

Asking "why" is consciousness' pivot foot. It allows for a virtual simulator to turn and think of varying options and what they portend. It allows the mind to revolve and go into different directions. As F. Scott Fitzgerald essayed in his book, *The Crack-Up*: "The test of a first-rate intelligence is the ability to hold two opposed ideas in the mind at the same time, and still retain the ability to function." Why is the mind's way of allowing a multiplicity of ideas to compete and hopefully function better because of it.

"Why" is similar to an all-purpose function key on your laptop computer which opens up programs that are otherwise hidden from display. But though asking why can be quite helpful in very specific situations (why does it rain in the winter and not the summer, for example?), it can also serve as unnecessary nuisance if its protestations cannot be adequately met. Perhaps this can help us better understand the wide gulf between religion and science. We have already admitted that for a virtual simulator to be highly effective it must be able to conjure up all sorts of imagined nonsense, provided it doesn't have to always act upon such in a real life situation. Science, though clearly built upon wild speculations and imaginings, is differentiated from religion because it measures its successes by actually "testing" its varying models with each other and placing them in real life contexts to determine which one holds up best under rigorous conditions. Science, in other words, attempts to falsify what consciousness conjures up so as to see which model best explains reality. And in so doing, it allows for a cataloging of both its successes and failures. In this way, science can indeed progress because it has a built-in tendency to eliminate less successful theoretic conjectures. Religion, on the other hand, tends to accept certain simulations above all others without resorting to any empirical verification and habitually substantiates such imaginary permutations as being beyond physical testing. In this way the virtual simulation protects its integrity and truth value by pointing to a transcendent arbiter and thereby foregoing any real world competition lest it be eliminated by such testing. Is it merely coincidence that there are tens of thousands of religions in the world each claiming exclusive truths, but nothing comparable in the world of science. There isn't a Japanese physics or a Tibetan physics or an American physics. There is just physics. What country you come from is secondary. Gravity is universal and doesn't have different acolytes claiming different revelations in different tongues. But which geographical region you come from in religion isn't secondary, but primary, since as every geographer knows the gods change when you go to different landscapes.

Virtual simulation can also be instrumental in helping us better understand why beliefs systems can be so powerful, even when such ideologies can be regarded as wrongheaded or backward. Any meaning system, provided it allows one enough purpose and drive to live another day, is better than none at all. As the script from the movie, Truth Lies, explains it:

Our brains didn't evolve to understand the universe but to literally "eat" it (as in survive the local ecosystem niche long enough to transmit one's genetic code). Thus, we have invariably conflated our appetite with truth. We don't know the truth; we simply know what it takes for us to live skillfully enough so that we don't get eaten too young. And if we live long enough, the redundancies (or spandrels or secondary effects) of our enlarged cerebral cortex allows us the freedom to ponder imponderables and impute upon those mysteries all sorts of silly nonsense with the added caveat that such idiocy will last provided it floats our boat to live another day. Let me rephrase that: nonsense evolved as an adaptive function of an enlarged brain because believing nonsense makes MORE sense (in terms of replicating strategies) than coming to grips with the random chaos from which the universe apparently arises. Tom Blake might put it this way, "nature is without sentiment, but those who FEEL sentiments have a better chance of surviving this horror show than those who do not." Why? Because any meaning is better than no meaning even if the universe ultimately is devoid of purpose. As Voltaire would say in my twisted way of paraphrasing him, "Man would have to invent God even if such a being didn't exist." We cannot live without purpose, even if that purpose is an adaptive fiction evolved over eons of time designed to blind those with such sentiments from the truth that nature has no such sentiment. Ah, too much truth and you cannot move. Too much reality and you become autistic. Which is another way of saying that Jack Nicholson was right all along: we cannot handle the truth. But the truth surely has a good way of handing us. It lies to us in order for us to live an extra day. Think about it. The truth is that truth lies

The idea that our brains could literally deceive us is now well established in neuroscience. Indeed, the brain's capacity for filling in objects that are not present is a vitally important component of how we navigate in our day to day world. As the abstract to "Perceptual filling-in from the edge of the blind spot" on Science Direct explains:

Looking at the world with one eye, we do not notice a scotoma in the receptor-free area of the visual field where the optic nerve leaves the eye. Rather we perceive the brightness, color, and texture of the adjacent area as if they were actually there. The mechanisms underlying this kind of perceptual filling-in remain controversial. To better understand these processes, we determined the minimum region around the blind spot that needs to be stimulated for filling-in by carefully mapping the blind spot and presenting individually fitted stimulus frames of different width around it. Uniform filling-in was observed with frame widths as narrow as 0.05° visual angle for color and 0.2° for texture. Filling-in was incomplete, when the frame was no longer contiguous with the blind spot border due to an eye movement. These results are consistent with the idea that perceptual filling-in of the blind spot depends on local processes generated at the physiological edge of the cortical representation.

The brain is forced to makes these "lying" choices to us as part of its mapping expertise. We are not seeing or hearing or smelling or feeling or touching the world "as it is," but rather as our brains "simulate" it for our evolutionary survival.

In this way philosophy is a multi-faceted procedure to encourage more simulations versus less. When Socrates axiomatically stated that an unexamined life wasn't worth living, he was arguing that we should bring to light more information about why and how we make the decisions we do. In his own inimitable way, Socrates was provoking the virtual simulator which is consciousness to start using its pivot foot (the "Why?") with more dexterity.

Evolutionary philosophy is in many ways similar to the Churchlands' concept of eliminative materialism. As the Neural Surfer website elaborates:

When we scientifically advanced in astronomy, medicine, and physics we replaced the old and outdated concepts of our mythic past with new and more accurate terminology which reflected our new found understanding of our body and the universe at large.

Thus, instead of talking about THOR, the Thunder God, we talked instead about electrical-magnetic currents. Thus, instead of talking about SPIRITS, as the causes of diseases, we talked about bacteria and viruses. Thus, instead of talking about tiny ghosts circulating throughout our anatomies pulling this or that muscle, we talked about a central nervous system.

In sum, we "eliminated" the gods or spirits in favor of more precise and accurate physiological explanations. Hence, the term: "eliminative materialism."

As a materialistic explanation evolves over time, it will either eliminate or reduce hitherto inexplicable phenomena down from the celestial region to the empirical arena. And in so doing, help us to better understand why certain events transpire in our body, in our mind, in our society, and in our world. Eliminative materialism is reason writ large.

The glitch, though, is that we have allowed eliminative materialism to change our thinking about almost everything EXCEPT ourselves. When it comes to understanding our own motivations, we have (as the Churchlands' point out) resorted more or less to "Folk Psychology," utilizing terms such as "desire", "motivation", "love", "anger", and "free will", to describe what we believe is happening within our own beings.

The problem with that is such terminology arises NOT from a robust neuroscientific understanding of our anatomies but rather arises from a centuries old MYTHIC/RELIGIOUS comprehension of our very consciousness. And that's the rub.

Where we have moved away from such religious goo speak in the fields of physics, astronomy, chemistry, and biology, in talking about ourselves we are still stuck in pre-rational modes of discourse. Where astronomy reflects the LATEST theories of the universe, where medicine reflects the LATEST theories of diseases, in talking about ourselves we tend to reflect ANCIENT theories

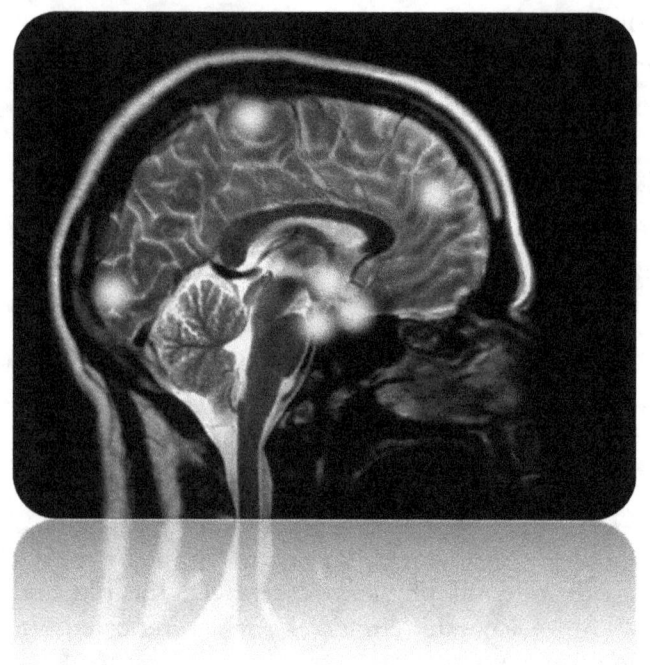

of human psychology. We resist knowing ourselves as MERELY this body, this brain, this material. As Patricia so astutely put it, "We don't want to be just three glorious pounds of meat." Well, according to eliminative materialism, that is PRECISELY what "we" are.

And in order to get a better understanding of human consciousness, neurophilosophy argues that we focus our attention on developing a more comprehensive analysis of the brain and how it "creates" self-reflective awareness. In so doing, we can then come up with a more neurally accurate way of describing what is going on within our own psyches (pun intended). Thus, instead of using the term "soul" we might instead use phase-specific words to describe the current state of awareness which are more neurologically correlated.

We have already done this slightly when it comes to headaches. Due to our increased attention to various pains and to the various drugs that are effective in treating them, we have become MORE aware of how to differentiate and thereby treat varying types of head pain. From Excedrin (very good for migraines because of the caffeine and aspirin combination) to Advil (very good for body and tooth aches). Hence, the neurophilosophical way to understand one's "soul" is to ground such ideas in the neural complex.

What may transpire, as Francis Crick suggests in his aptly titled book *The Astonishing Hypothesis*, is that the soul will disappear. Why? Because there really is NO soul. We are rather a bundle of neurons and nerve endings tied to together in a huge neural complex that gives rises to consciousness. There is nothing META (beyond) physical about us. We ARE physical. And that very insight will lead to a reinterpretation of who we are and why we are and how we are.'

If consciousness does indeed serve as a virtual simulator with an amplified probability feedback loop, then it should come as no surprise that one of the more promising theories arising from neuroscience concerning how the brain works is based upon Bayesian probability theory.

From the New Scientist:

Neuroscientist Karl Friston and his colleagues have proposed a mathematical law that some are claiming is the nearest thing yet to a grand unified theory of the brain. From this single law, Friston's group claims to be able to explain almost everything about our grey matter.

Friston's ideas build on an existing theory known as the "Bayesian brain", which conceptualises the brain as a probability machine that constantly makes predictions about the world and then updates them based on what it senses.

The idea was born in 1983, when Geoffrey Hinton of the University of Toronto in Canada and Terry Sejnowski, then at Johns Hopkins University in Baltimore, Maryland, suggested that the brain could be seen as a machine that makes decisions based on the uncertainties of the outside world. In the 1990s, other researchers proposed that the brain represents knowledge of the world in terms of probabilities. Instead of estimating the distance to an object as a number, for instance, the brain would treat it as a range of possible values, some more likely than others.

A crucial element of the approach is that the probabilities are based on experience, but they change when relevant new information, such as visual information about the object's location, becomes available. "The brain is an inferential agent, optimising its models of what's going on at this moment and in the future," says Friston. In other words, the brain runs on Bayesian probability. Named after the 18th-century mathematician Thomas Bayes, this is a systematic way of calculating how the likelihood of an event changes as new information comes to light

Over the past decade, neuroscientists have found that real brains seem to work in this way. In perception and learning experiments, for example, people tend to make estimates - of the location or speed of a moving object, say - in a way that fits with Bayesian probability theory. There's also evidence that the brain makes internal predictions and updates them in a Bayesian manner. When you listen to someone talking, for example, your brain isn't simply receiving information, it also predicts what it expects to hear and constantly revises its predictions based on what information comes next. These predictions strongly influence what you actually hear, allowing you, for instance, to make sense of distorted or partially obscured speech.

In fact, making predictions and re-evaluating them seems to be a universal feature of the brain. At all times your brain is weighing its inputs and comparing them with internal predictions in order to make sense of the world. "It's a general computational principle that can explain how the brain handles problems ranging from low-level perception to high-level cognition," says Alex Pouget, a computational neuroscientist at the University of Rochester in New York (Trends in *Neurosciences*, vol 27, p 712).

Friston developed the free-energy principle to explain perception, but he now thinks it can be generalised to other kinds of brain processes as well. He claims that everything the brain does is designed to minimise free energy or prediction error (*Synthese*, vol 159, p 417). "In short, everything that can change in the brain will change to suppress prediction errors, from the firing of neurons to the wiring between them, and from the movements of our eyes to the choices we make in daily life," he says.

Applying Friston's grand unified theory of the brain to consciousness, it can be argued that while minimizing prediction errors is elemental and particularly relevant for any modeling system working in the real world of life and death, the most marked feature of second order awareness is its dissociation from such real world onslaughts. The fact that our awareness can be freed from the present struggle for existence is perhaps why it is so useful in orienting us to future occasions. To make a crude analogy here, a drowning man doesn't have enough "free" time or energy to do philosophy. His first nature instincts must take over and his reptilian brain survival tools must kick into high gear. However, if the drowning man is saved and is allowed enough leisure time to reflect (another word for simulate) upon what happened to him, he may be able to play out an array of options for escaping just such a dilemma the next time it might happen to him. But this presupposes a surplus of both physical and mental energy which is not already obliged or preoccupied with any over-arching present conundrums.

In light of this necessity for leisure, it can be argued that philosophy can only be practiced in earnest when there is enough free time and energy. A pilot who is flying under enemy fire doesn't have the option to go and ruminate in his mock-up flight simulator. Likewise, philosophy can only arise in a sustained fashion in a culture which has ample time on its hand.

Going back to our basketball analogy, the center can't use his or her pivot foot if he or she is too closely guarded. The same may also be true to some extent with the mental use of our pivotal "why." If conditions are too severe, we won't have the freedom to turn our thinking around or rotate our concepts in new directions. If time is of the essence, philosophy is not. Indeed, one can even propose a syllogism on the basis of this questionable claim. If you are doing philosophy for any measurable period of time, it can be taken as conclusive proof that you have too much of it. This, of course, is not to disparage philosophical inquiry but only to underline how an evolutionary approach to the subject uncovers the basis for why such an endeavor would emerge in the first place and why it can only be practiced under certain optimal conditions.

Recommended Readings

Ten Annotated Books on Evolution Theory

Darwin's Dangerous Idea: Evolution and the Meanings of Life *by Daniel Dennett*

Daniel Dennett breaks his book up into three sections: the first deals with exactly what Darwin's dangerous idea is; the second section more or less examines the biology of evolution; and third part looks at how Darwinian evolution has transformed our understanding of who and what we are.

Section One:

What is Darwin's dangerous idea? Darwin's idea is the single best idea anyone has ever had, argues Dennett. It is also the most dangerous one. What he means by this is that it burns, like a "Universal Acid," through any misconceptions we have about nature. Special Creation is burned away; the Cosmic Pyramid of God, Mind, Design Order, etc. is annihilated; Plato's essentialism is destroyed; Locke's primacy of Mind is no more. Darwin single handedly demystified the world with his reductionism and usurped all of our traditional understandings in one swoop. He replaced a "skyhook" designer with an algorithmic "crane." And, yes, without a designer, Dennett quipped, there can still be design via this algorithmic process in nature and the statistical probability of design arising after billions of years of hit and miss tries. This is where intelligent design theorists get it wrong: there can be design without a designer. Moreover, this is where "greedy reductionists," who altogether dismiss design apparent all around us, also get it wrong.

Science offers us a totally new perspective of the world and who and what we are, and, hence, science and philosophy forever are intertwined. As with the Copernican Revolution where the shot was heard around the world, Darwin's dangerous idea is still making its way around the world. It took centuries before everyone accepted Copernicus' heliocentric model and it may take the same time or more for the Darwinian Revolution to dominate.

Should we fear this dangerous idea in anyway, asks Dennett? Absolutely not! We just need to grow up, he says, and embrace the underlying beauty of it. Some philosophers have accused Dennett of being the very thing he criticizes, a greedy reductionist. But this seems to be an unfair assessment of Dennett, especially in light of the fact that he yearns to see the magnificence within the natural world via evolution.

Section Two:

I loved the opening quote of this section: "Nothing make sense except in light of evolution" (Theodosius Dobzhansky). This quote sums up the theme of section two. The "Laws of the Game of Life" (i.e., biology which he calls engineering) can only be understood in terms of evolution. The laws and regularities we witness in nature really rely on blind, meaningless chaos. There is no Universal Mind or Wizard of Oz behind the curtain pulling the strings. Instead, the world we live in today is a result of what Crick called "frozen accidents." Sure, Paley witnessed design and he was right to. But design is the accumulation of billions of years of a mindless, algorithmic process. Nietzsche's Eternal Recurrence theme is placed in this section to stress the meaninglessness of life and the only meaning we find is that which we create ourselves.

I especially appreciate his discussion of "intellectual tennis" in section two. Defenders of Special Creation want to play "intellectual tennis" with the net down on their side when serving (referring to not following rules and not offering evidence) but up for the opponent. This is certainly unfair in the field of science.

Finally, Dennett exerts a lot of energy challenging Gould's perspective of evolution and science. He refers to Gould as the "boy who cried wolf" and even had references to him as a "bully." Why is Dennett so taken back by Gould? Well, quite simply, he accuses him of misrepresenting Darwinian evolution. While Gould brilliantly contributed the idea of spandrels to evolutionary theory, his resistance to gradualism, Dennett contends, is off putting. Other controversies, including Teilhard's Omega Point theory, are shot down in this part of the book.

Section Three:

The third section of the book starts with meme theory. Dawkins' memetic understanding of cultural evolution gets the thumbs up from Dennett. Language itself plays a crucial "crane" role in the development of the human mind, though Chomsky has resisted the interplay between Darwinian evolution and linguistics. Instead, Chomsky, along with Searle, contends Dennett, understands the mind more as a "skyhook" than a "crane," especially in both of their rejection of Artificial Intelligence as modeling human intelligence. Moreover, Penrose's meme of Godel's Theorem as proof against AI needs to be "extinguished," says Dennett.

Having highlighted the power of Darwin's dangerous idea still further, Dennett turns his attention to morality and evolution. Are sociobiologists being greedy reductionists by reducing morality to a product of evolution? Certainly, he petitions, we need to understand ethics along Darwinian lines but perhaps not to the level the greedy reductionist would take us. Here Dennett argues that we are not set creatures but that we have the "mind-tools to design and re-design ourselves" and even to re-design moral codes themselves.

In the conclusion of the book Dennett suggests that upon closer inspection Darwin's dangerous idea is not a "wolf in sheep's clothing," but a "friend mistaken as an enemy." Here he is referring to the famous story Beauty and the Beast. Instead of being a terrifying beast, Darwin's dangerous idea is really "a friend of Beauty, and indeed quite beautiful in its own right."

The Making of the Fittest *by Sean B. Carroll*

In this country we use DNA every day. We use it in the court system and crime scenes; we use it in paternity tests; etc. But, the author asks, why do fifty percent of Americans take issue with implications of DNA in understanding evolution? Since we have mapped out the entire human genome, our comprehension of DNA in the natural world has grown "40, 000 fold." Genomics, the study of DNA, simply overwhelms us with solid evidence of evolution, asserts Carroll.

The author goes to great length to demonstrate through several fascination examples evolution in the natural world. The icefish, a bloodless vertebrate, evolved as it did with the loss of hemoglobin in order to survive freezing temperatures of the Antarctic. Thomas Huxley's "natural selection" (he was the first to coin the phrase) was certainly at work here. Carroll also offers us an example of how DNA itself clearly illustrates evolution. Fossil genes (namely, DNA left over from our ancestors) are found in all species throughout the world.

Referring to his book as "genocentric" (i.e., concentrating mostly on DNA), Carroll wrote *The Making of the Fittest* for the average layperson with the objective to explain how evolution works. The topic of evolution and the evidence of it need not be an erudite subject reserved for scientists, he says.

Darwin's DNA

Carroll addresses several questions in his effort to clarify to the public what evolution is. For instance, he asks how often do mutations occur in nature? Are they the norm? Carroll walks us through the statistics on this. He explains that among 7 million DNA letters we experience about 175 mutations. But why don't we see mutations on the human species all the time? Most of the mutations occur on what is called "junk DNA," and this is not purged unless its effects are negative. Also, since we carry two copies of genes sometimes mutations do not show up on one gene. Despite all of this, the author refers to us as "all mutants." We are a product of these mutations. It is important to note that natural selection "rejects the changes that are harmful, favors changes that are beneficial, and is blind to changes that are neutral."

What is most fascinating is how many genes we have compared with other species. Homo sapiens possess around 25,000 genes. Many plants, mice and fish carry the same number. Even a worm has 20,000 genes. A fruit fly has two times the genes as yeast and a human has about two times the genes of a fruit fly. As complicated as humans are, why do we not possess more genes than other species? Interestingly, many of our genes are in fact redundant.

In this book we find out that evolution repeats itself many times over. What a great concept! Carroll gives several examples of how evolution does this. For instance, different species develop similar toxins or similar flippers or similar vision. Statistics illustrate that over the vast course of time and having had numerous offspring, "identical or equivalent mutations will arise repeatedly by chance."

The genome of any species contains "a record of the history of life." It holds information about not only its own species but also all the preceding species to the beginning of life on earth. There are about 500 immortal genes (or fossil genes) that all seem to share. The retaining of genes and losing of genes over time demonstrate that there is no concept of progress in evolution. Natural selection is not about design or intent. We are not better adapted than other species of the past. We are just different. William Paley's design arguments falters when one considers, as Darwin did, that vast amounts of time of gradual and sometimes rapid change explains the rich variety of life forms we see today without invoking the notion of a designer.

The anti-evolution movement is lead by the religious right who do not understand genetics and the power of DNA research in proving evolution at play. Creationists do not understand the connection between randomness at the DNA level and natural selection. The author takes great effort to clarify that natural selection does not entail randomness at all. As he says it, "selection, which is not random, determines what chance occurrences are retained." Cumulative selection over eons of time produce the rich tapestry of the organic matter we see today. Evolution is simply the interplay of "chance and necessity." Moreover, Creationists do not understand what a scientific theory entails. It is not an educated guess but "a well substantiated explanation of some aspect of the natural work that can incorporate facts, laws, inferences, and tested hypothesis." The evidence for evolution is overwhelming and it cannot be argued away. There are indeed religious minded people, such as Pope John Paul and the scientist Kenneth Miller, who embrace evolution as factual.

Quoting Peter Medawar, "the alternative to thinking in evolutionary terms is not to think at all." This, Carroll purports, is something we as a species cannot afford. In fact, Carroll ends his book with a chapter on how not understanding evolution has led and will continue to lead to enormous gaffes on the part of humans in nature. Our actions have resulted in destroying balanced ecosystems but it is our responsibility to understand and respect the dynamics of nature. Over-fishing has depleted the waters of certain species of fish; pollution has contributed to the warming of the oceans; many animals are near extinction at our hands. There is no more time to debate the fact of evolution as we destroy elements of this planet. Two centuries of detailed science has moved evolution from the hypothesis stage to the scientific theory stage (that which is supported by a variety of facts, not the least of which is DNA). It is time to wake up and take responsibility, even if out of self-interest, as he put it.

Dawkins vs. Gould: Survival of the Fittest *by Kim Stefelny*

This book examines the commonalities and, more importantly, major differences between Dawkins and Gould on the theory of evolution.

Let us begin by looking at what they have in common:

Evolution occurred 4 billion years ago.

The process was natural, not divine.

Chance plays a role.

There is no aim, no purpose.

There is nothing inevitable about humans.

Variation in populations leads to varying reproductive chances.

Natural Selection is essential for evolution: the more fit in the environment has more descendents.

Natural Selection for humans works slowly over generations but fast for short term organisms.

Large, random changes are mostly disasters since change must be gradual and cumulative; but on occasion and very rarely there is one big mutation. Overall, though, evolution is generally a long series of small changes.

Differences between Dawkins and Gould abound as well; both represent totally different traditions in evolutionary biology. The focus of this book is on these main differences:

Dawkins vs. Gould:

Dawkins: Dawkins looks at ethology and problems of adaptation.

Gould: Gould, on the other hand, is a paleontologist who studies fossils; he sees himself as a historian of life.

Dawkins vs. Gould:

Dawkins: Dawkins focuses on evolution as a history of gene lineages where there are long periods of small changes. Genes replicate almost identical to predecessor. Struggle of evolution is struggle of genes to replicate. Yes, this is reductionistic but genes make the organism.

Gould: Gould instead concentrates on extinction and not persistence of gene lineages. He looks at the basic blueprint of animals noting that there are no fundamental new inventions. Gould argues that chance is more important and genes less important. Luck plays a role he says and not just fitness. Catastrophes happen, he points out, and the lucky survive, not the fit.

Dawkins vs. Gould:

Dawkins: Dawkins carries a different attitude about science than Gould. He is the son of enlightenment who looks at science as the best way to gain knowledge. Science is complete and beautiful, he says; it is open to revision with new evidence and new ideas. Finally, Dawkins argues that science is not to be understood as "the dominate ideology of the time but engine of objective knowledge."

Gould: Gould, on the other hand, views science as not complete; there are social influences on scientific views (his book *Mismeasure of Man* illustrated this). Basically, he argues that the ideology of the time impacts science.

Dawkins vs. Gould:

Dawkins: Dawkins' views on religion vary from Gould's; he is an atheist and sees religion with no authority on values. Religion, he asserts, is a meme, and a destructive one at that. He goes as far as to call religion a virus that we need to get rid of.

Gould: Gould argues that science and religion are different domains. Science does not study moral claims where religion does, he states. Gould does not address theism questions though and seems to avoid them altogether. Also, Gould does not support postmodern relativism.

Dawkins vs. Gould:

Dawkins: Dawkins supports evolutionary psychology – the using of evolution to understand our behavior; he argues that this is a liberating idea since it liberates us from religion.

Gould: Gould does not like evolutionary psychology / socio-biology; he argues that it is a dangerous idea that can justify heinous acts like rape.

Dawkins vs. Gould:

Dawkins: Dawkins argues for natural selection acting on genes/gene lineages Adaptation, quips Dawkins, evolves slowly in small steps with occasional large changes; overtime there is a large range of variation possible.

Gould: One the other end, Gould focuses on natural selection on organisms and not genes per se; Gould argues that evolutionary possibilities are constrained since they are basically entrenched.

Overall, this was a fascinating read on how two respected scientists in the field of evolution can so drastically disagree. What this illustrates, to me, is that science is not a dogmatic discipline but a healthy one full of lively debates that keep inching us along to a more accurate understand of how the world works.

| Darwin's DNA

Why Darwin Matters *by Michael Shermer*

Shermer argues, rightfully so, that the ID movement is mainly a front for creationism. He examines the history of this topic from the Scope's trial to the recent Dover trial. Overall, this reading is a superb source to understand the problems with the ID movement.

There are a variety of problems with the ID movement. Here is a list of some of the major problems Shermer addresses:

1. The ID movement is a front for a religious position.

2. The Second Law of Thermodynamics is misunderstood.

3. Either – or – Fallacy (accept God as Designer or accept total chaos) is prevalent in the ID movement.

4. The burden of proof falls on claimant and extraordinary claims require tons of proof.

5. The ID movement argues that randomness does not produce the world or apparent design in nature; however, they do not understand that randomness actually builds on itself and so quite significant results can happen rapidly.

6. There are multiple creation accounts in the world and there should not be equal time.

Shermer addresses key questions of evolution. Here are some important queries:

1. Can you believe in god and evolution? His answer: Yes. He uses the NOMA argument of Gould.

2. Should the conservative Christians embrace evolution? Yes. Science need not contradict religion. He uses the Pope John Paul example to show that science and religion can work together.

3. Why do we see design? Because we are designed to see a Designer!

4. Why is there a shortage of fossils? It is rare to find a fossil since animals in past were eaten as a general rule; however, there are plenty of intermediate forms available today.

Darwin's DNA

The author highlights the evidence of evolution. The following are some examples of the evidence he covers:

1. Dawkins' *Ancestor's Tale* is one of the best compilations of evidence.

2. The concept of scientific "theory" is misunderstood; a theory really it is a fact supported by all of the natural sciences and evolution has earned this title.

3. Grant's work on finches, a twenty year study, verifies Darwin's data.

Shermer discusses strange designs in nature that evolution explains:

1. Males have a form of a uterus off the prostate gland.

2. Males also have nipples.

3. Just as apes have a 13th rib so do 8% of humans.

4. Humans have a tailbone.

5. We have an appendix for digesting cellulose (primal diet was mostly vegetarian).

6. Strangely, we have goose bumps and these resemble the raising of animal fur.

In the Epilogue, Shermer's final comments match Chet Raymo's work that the scientific world view does not diminish spirituality or beauty in the world. Science can foster a feeling of spirituality as one feels connected to an infinite universe and "moved to tears" from scientific revelations. Attempting to illustrate how amazing this multi-universe is, Shermer coins a new term he calls "sciensuality, meaning a feeling of awe, humility and sensuality of discovery." Science tells us the story about "who we are, where we come from and where we are going." Thus, he says, Darwin indeed matters.

In the final section of the book, the Appendix, Shermer discusses Eugenie Scott's argument that there are at least 8 different religious views on the creation-evolution issue; these are the following:

1. Young earth creationism

2. Old earth creationism

3. Gap creationism

| Darwin's DNA

4. Day-age creationism

5. Progressive creationism

6. ID creationism

7. Evolutionist creationism

8. Theistic evolution

Darwin Loves You *by George Levine*

The title of this book, *Darwin Loves You*, is a take-off of the "Jesus loves you" bumper sticker. But instead of a Christian premise, Levine illustrates that Darwin's research radically enchants the world with a sense of wonder and awe in nature. Much like Chet Raymo's thesis in *Skeptic and True Believers*, Levine illustrates that Darwin, and science itself, does not in any way de-mystify the world. Rather, scientific research such as Darwin's is able to mystify the world as we are left dumbfounded by the awesome revelations it reveals. Specifically, the *Origin of Species* gives us an appreciation for how amazing nature is.

Throughout the text the author counters Max Weber's position (as well as William James') that science, following the rational school of thought, disenchants the world and creates a world void of meaning and value. Instead, Levine makes the case that Darwin serves as a model for what Shermer calls "sciensuality." While religion tends to devalue the world, Darwin, argues Levine, offers us a new view. Thus, Weber has it backwards: science is not the one that demystifies the world but religion does. Science offers us a healthy view by explaining the world naturalistically with a deep connection to nature.

Darwin ennobles the earth, he explains clearly, and Weber was simply wrong. In fact, Darwin, through his work, offers "spiritual, cultural, and ethical value." Yet, Levine is talking about a non-theistic enchantment, a wonderment experienced by understanding the world materialistically and naturalistically. For Darwin, matter was not inert or meaningless but alive and vibrant.

Even Darwin's son, William, contended that for his father religion was in nature, and this insight offered religious feelings of awe. More accurately put, science can offer us life with "moments of enchantment." But the author was quick to clarify that moments of enchantment was not necessarily an enchanted life, just glimpse of the magnificence of nature.

Though we live in a wonderful world full of awe, beauty and mystery, it can still be a scary and dangerous place filled with bombs, rapes, slaughters, pain, as John S. Mill pointed out. While science does not justify pain, through science one can explain it. More importantly, the enchantment idea can offer us a reprieve from the pain. For instance, understanding how the eye works via evolution is magnificent and may lessen a feeling of being overpowered by nature and its unexpected turns.

This insight also parallels the "love the earth, our home" theme of Nietzsche when he petitions us to love the world despite the hardships and pains because this earth is our home and all we have and really know. The Ubermensch has the courage to face the world as it is and emerge transformed.

For the author sociobiology and evolutionary psychology are too reductionistic for his taste and contribute, he says, to disenchantment of humans. Just like the 19th century Victorians, scientists such as Pinker, Dawkins and Dennett fit in the reductionistic, positivistic camp. Darwin, while he embraces reductionism, argues Levine, does not fit in the exact same school of thought. Instead, he practices what the author calls "good reductionism" and not "greedy reductionism" or bad science. Levine even places Edward O. Wilson, the author of *Consilience,* in this radical reductionistic group for explaining humans via genetics. Levine concedes that Wilson has some romantic elements in his consilience theory but enchantment is lost when everything in the natural world is reduced to scientific law.

According to Levine, the affect of awe that Dennett and others miss is found throughout the *Origins of Species,* almost with "exclamation points." (Note: having read Dennett's work I am not sure this is a fair assessment of him.) What Darwin saw that amazed him, that offered richness and wonder in nature, was the incredible understanding that we are all related, that all of nature is connected. Humans lose their anthropocentric position but in place of it we become part of the greater whole. Epictetus' message of interconnectedness dominates.

There are two more key points that Levine addresses: Darwin was affiliated with biases of his cultural times but in no way does this diminish the validity of his work. And the author looks at the misuse of Darwinian theory to justify social, political aims (e.g., Social Darwianism). He defends Darwin despite abuses of his theory by others because an "is" does not imply an "ought," as Hume brilliantly posited.

The Reluctant Mr. Darwin *by David Quanmmen*

A wonderful and insightful biography on the life of Darwin, this book focuses on Darwin's life post the Beagle adventure and it makes clear his internal struggle to publish his masterpiece, *Origins of Species*. Several factors contributed to the delay of publishing it: major physical ailments; family tragedies like the death of two children; and, probably most significant, a fear of society's backlash.

In terms of Darwin's religious background, he was raised both secular and a liberal Christian. The reason he first attended theology school to become a parson was because such a profession allowed him, after the Sunday obligations, the rest of the week to pursue a life as a naturalist. Darwin eventually rejected this career as his theism waned. His theistic perspective was followed by deism and then eventual secular materialism. However, elements of theism are still found in the *Origin of Species*, published in 1859, when he refers to the Creator. Darwin even quotes William Whewell's "Divine Power" idea as the creator of "general laws."

But is Darwin truly a theist? No. He sees too many questions in nature that forces him to ask: why would a benevolent god create so much cruelty and chaos in nature? Examples of nature's frenzy include the laying of wasp eggs inside a caterpillar so that the larva could eat the host upon hatching. Even the death (most likely from tuberculosis) of his favorite and oldest child, Annie, shows the "problem of evil" that Darwin never seems to reconcile.

The author brilliantly argues that Darwin does not really challenge the existence of god as he does the godliness of man. By this saying this, Darwin is concentrating on the interconnectedness of all of nature, of all life forms, and the humans do not stand apart from or above it. Darwin refers to this interconnection as the "grandeur of life."

This grandeur is clearly illustrated when we compare the human genome with that of a mouse's. The 30,000 genes that we both have are 99% direct "counterparts" (meaning very similar but not identical as a chimp and a humans are). Life is from one original source. Embryology shows that we have signs in the embryo state of the progenitor before (e.g., tail, gills, body hair, etc.)

Biogeography, a field that deals with the spread of animal and plant distribution on the planet, was of particular interest to Darwin. He was fascinated with how life forms can be spread across the world. Seeds survive in animal feces and are passed to other areas. Clams hook on to water bettle's legs and as they catch a ride develop habitat elsewhere. Interestingly, the last clam/bettle specimen Darwin was given came from the grandfather of Francis Crick.

What was Wallace's role in this field? He sent Darwin a manuscript detailing his evolution ideas that included transmutation and natural selection (the latter was not coined by Wallace) prior to the publishing of *Origin of Species*. Darwin, along with his scientist friends, decided to give Wallace co-recognition in a paper at a science lecture on the topic. Darwin then was given the impetus to overcome his prior reluctance, fearing that someone would gain priority in the field. The *Origins of the Species* shortly followed, selling out of first edition copies on the first day of sales. Both scholars argued for gradualism and against Lamarkianism, but later on Wallace, a supporter of spiritualism, argued that natural selection could not account for the development of the human brain. Darwin refers to Wallace's slip into spiritualism as intellectual suicide.

Overall, this book was quite an enjoyable biography. It offers insights to Darwin's personal life that make his scientific endeavors even more impressive.

The Ancestor's Tale *by Richard Dawkins*

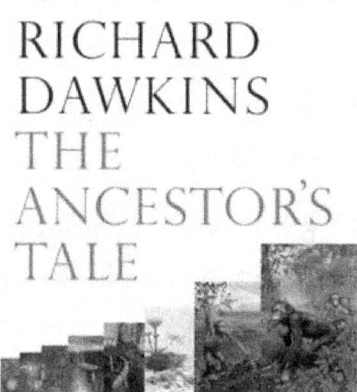

Dawkins begins his tale by explaining that humans are pattern, meaning seeking creatures. Being the pattern seeking creatures that we are, humans look at the history of the world and believe that it was fine-tuned to bring humanity into existence (basically, this is the anthropic principle). He refers to this as the "conceit of hindsight." Our conceit is the belief that the past works to deliver the present, namely us, and so humans are the final goal in evolutionary history.

But is evolution progressive? No, argues Dawkins, not at all. We are wrong to think that our predecessors were transitional beings and a halfway mark to us. The idea of us being "more evolved," however, comes from an incorrect, human centric perspective. Dawkins humorously points out that if elephants could write history they would see everything as leading up to the development of the trunk.

This text takes a different approach to understanding evolutionary theory. Instead of starting with the earliest forms and showing the diversity of life, Dawkins goes backwards, beginning with humans, to show the unity of life. This backward approach traces humans back to a shared ancestor with the apes, more than 18 million years ago and eventually back to the progenitor of all, the bacterium. The point of a shared ancestor, the moment of a rendezvous of a last common ancestor, is called a "concestor." Is there a concestor of all life forms? Yes, since we share a common genetic code with all of life on this planet. The Grand Concestor of all life forms, bacterial fossils, dates 3.5 billion years ago.

Dawkins adventure begins with the Homo sapiens as he pilgrimages from the present to the past common ancestor. On our journey the first pilgrims we meet are the chimps 5 million years ago. This is followed by the gorillas, then the orangutans, to the gibbons, to the old world monkeys, then to other mammals and eventually to the first form in the sea.

There are only 40 rendezvous points in the human pilgrimage to the origin of life. In other words, there are 40 steps or 40 concestors in our history to the first life form. Concestor number 39 is the grand ancestor of all life forms. These 40 natural milestones is a literary allusion to Chaucer's Canterbury reference of a pilgrim's tale.

Dawkins points out three main methods to understanding evolution: archeology (hard relics such as fossils, etc.); renewed relics such as DNA; triangulation, that is, triangulating an ancestor by comparing two surviving descendents. Let us look at each of these briefly:

1. Fossils: fossils demonstrate evolution but even if there are gaps in the fossil records there is overwhelming evidence of evolution; fossils are just a bonus.

2. DNA: the alphabet of DNA is a like a writing system that records and copies. DNA changes very slowly so the record is woven into all plants and animals. Four letters allows for 64 limited words. There is a 64 word DNA dictionary that is universal and unchanging. DNA, argues Dawkins, is the genetic book of the dead, a chronicle of the past.

3. Triangulation: this is the comparison of DNA sequences of species and a look for shared DNA; for instance, comparing human and bacteria DNA to find overlap and resemblances. An analogy would be a similarity of languages like German and Dutch. But human and chimp DNA is so similar it is like English spoken in two slightly different accents but still English nonetheless. The analogy continues when you compare English and Japanese---there are no two organisms that can compare to the great differences between these two languages, not even humans and bacteria. All of this shows a definite family tree among organisms.

The actual pilgrimage now begins. If we look at the first tens of thousands of years (say 50,000 years ago) there really is no genetic difference than we are today. There are two major cultural advances in the last 50,000 years: agricultural revolution (farmer) and Cro-Magnon's tale (the flowering of the human mind). The farmer's tale was an agro-revolution that began 10,000 years ago at the end of the last ice age and is called the new stone age or Neolithic. There is a transition from the hunter-gatherer and a new idea of home. At this time large populations are supported but diseases are now a threat; domestication, farming and urban life are central. This period was preceded by what Dawkins calls the cro-magnon tale. It began 40,000 years ago as the hunter-gather society was dominant with music, figurines, graves, paintings, carvings, etc. There was no longer a million year stagnation but the flowering of consciousness in the Homo sapiens; some even attribute this to the origin of language.

Dawkins speaks a great deal about concestor 0. But what is concestor 0? It is the most recent ancestor of all surviving humans. MRCA refers to the most recent common ancestor. Mitochondrial DNA MRCA is called mitochondrial Eve; for males the MRCA is called y-chromosome Adam. Adam and Eve could actually be separated by tens of thousands of years; Eve most likely preceded Adam; she was probably 140,000 years ago and Adam 60,000 years ago.

Rendezvous 0 refers to tens of thousands of years ago to hundreds of thousand at most. And Rendezvous 1 refers to the fork between humans and chimps millions of years ago. Dawkins makes an interesting claim that you may be more related genetically to a chimp than to some humans. One big difference between humans and chimps, though, is the FOXP2 gene. This FOXP2 gene is the gene that allows humans to have language. Since chimps lack this gene they also lack language as we have.

An interesting question is: are we related to the Neanderthals? And is there interbreeding between Neanderthal and humans? Most likely the answer to both of these is no since mitochondrial Eve does not match them; few if any genes of ours come from them.

Homo ergaster or Homo erectus lived one million years ago. And actually they were no more erect than predecessor Homo habilis. They are altogether a different species than us. Moreover, it is not conclusive if they used fires but there is some evidence of campfires; tools were used and we are not sure if they had developed some form of language.

Homo habilis lived two million years ago and are referred to as the Habilines. What is unique about them? These handymen from Africa had larger brains; 750 cc brain was significant since the larger brain now fits the homo category. The Autralopithecus was its predecessor.

The brain size of the Homo sapiens is 6 times larger than it should be for a typical mammal of equivalent size. The Homo habilis is about 4.5 times bigger than it should be and a Homo erectus is about 4 times larger. For a chimp its brain is about 2 times larger than it should be for a mammal that size. What accounts for the EQ (encephalisation quotient) of humans? There are various theories that Dawkins considers. Certainly, it is our brain size that makes us a bit unique but in no way superior.

Australopithecus afraenis lived 3.2 million years ago; Lucy, a bipedal being, is an example of this found in Ethiopia in 1974. Why the rise to bipedal? There are numerous theories; these include: having extra height to see; squat feeding to turn over rocks looking for insects; freeing of hands to carry food; having less sun on the body. Dawkins argues that there may be several factors that account for it. Little Foot is a 4 million year old bipedal fossil whose toes also suggests tree climbing too.

Overall, the *Ancestor's Tale* is an essential read to further one's knowledge on the subject of evolution. Michael Shermer, in *Why Darwin Matters*, goes as far as to call the book one of the greatest compilations on evolution ever. And certainly I agree.

| Darwin's DNA

The Human Genome Sourcebook *by Tara Acharya and Neeraja Sankaran*

This textbook is essentially a summary of the 1990s Human Genome Project, also referred to as the "Book of Life," or the "blueprint" or "recipe book." The recipe is the genome, derived from the words gene and chromosome.

While the *Human Genome Sourcebook* was written with the general reader in mind, it is still scientifically challenging. The authors refer to their work as a reference volume and certainly that is what it is. This reference text begins with a historical overview of heredity beginning with Lamark's, Darwin's and Mendel's research. It then continues to give a history of genetics. In the early 1900s William Bateson coined the term "genetics," referring to genesis of traits. Some of the first genetic research on chromosomes began with Thomas Morgan's work on fruit flies at Columbia University in the early 1900s.

Here is some of the information catalogued within the text:

DNA inside each cell, if stretched out, would be over 6 feet long.

Chromosomes vary with each species: while humans have 23 pairs, a peanut has 20, a dog has 39 and a sugar cane has 40 pairs.

The entire human genome consists of 3.1 billion base pairs.

The text highlights the numerous ethical concerns regarding genetic research, namely genetic profiling and eugenics.

A large portion of this book is dedicated to defining genetic terms and concepts from genotype to mitochondrial DNA. An understanding of chromosomes and genetic diseases follows the term section. Cancer, diabetes, lupus and the like are genetically explained.

While reading the text my focus was on evolution: correctly explained was the idea that apes or monkeys did not give rise to humans, but that we share a common ancestor with them. The authors clarified that evolution occurs at a slower rate the larger the genome size. Our 3 billion base pairs results in a slower rate of change than species with a smaller number of base pairs.

While not entertaining per se, one learns a lot within the pages and so it was, overall, a worthwhile read.

DNA: The Secret of Life *by James Watson*

James Watson, co-founder of the structure of the DNA molecule, begins with a discussion of the origins of genetics. Interestingly, he points out that it really began with the Greeks, including Hippocrates. Pangenesis (the transfer of miniature parts of the body via sperm/egg) was a popular idea of the time. Even Darwin entertained pangenesis to explain inheritance, since he was unaware of Mendel's breakthrough research on genetics. Pangenesis idea was finally decimated when amputated tailless mice kept producing offspring with tails. As in the *Human Genome Sourcebook*, Morgan's research on fruit flies was discussed; fruit flies make a great specimen since a new generation only takes ten days and a female lays hundreds of eggs.

Francis Galton in the 19th century introduced eugenics, meaning "good in birth." Even George Bernard Shaw referred to eugenics as that which "can save our civilization." There were two approaches: Galton's positive eugenics that encouraged gifted people to procreation; and negative eugenics that tried to prevent those viewed as not superior from reproducing. The IQ test in the early 20th century contributed to a fear that bad genes were entering American soil. Henry Goddard argued that immigrants were the cause of the downward spiral of American intelligence. Theodore Roosevelt bought the argument. Sadly, racism, while not inherent to eugenics, became connected to it. Hitler became part of the movement as well.

Watson retells the story of how he and Crick discovered the structure of the DNA molecule. One competitor scientist was Rosalind Franklin, who died at 37 of ovarian cancer. Feb. 28th, 1953 marks the day the discovery was made. Vitalism was put to an end at this time. Soon after the discovery of the structure of the DNA molecule, evidence of genetic mutation was conclusively demonstrated. Usually A and T matched up and C and G matched up; but on occasion there was a genetic mix up and the sequence changed.

Several interesting chapters, such as the ones on biotechnology and genetically modified agriculture, followed the explanation of the DNA structure. The human genome project receives special attention.

My favorite section was chapter nine on DNA and evolution where our human past is examined. Homo neanderthalensis can be dated 30,000 years ago and are described as a different yet similar species to our own. They had slightly larger brains and there is evidence they buried the dead and may have believed in an afterlife. DNA solved the mystery of our relations to them. When one scientist working on the DNA sequencing referred to the excitement as "something starting to crawl up my spine" I was reminded of Chet Raymo's argument that good science produces a "shudder up the spine."

Strangely, it turns out that we were more closely related to Neanderthal than to chimpanzees and so Neanderthals are considered part of the human tree (a branch but not part of the same limb). However, we do not see Neanderthal genes mixed with ours so interbreeding is suspect. While we are genetically different than the Neanderthal, this research indicates how connected we are overall with the natural world.

One question that caught my attention was: at what point did the human lineage separate from the chimps and gorillas? Mary-Claire King's research indicated it occurred no more than 5 million years ago. And she also showed that humans and chimps differed genetically only by 1%. Gorillas and chimps differ by 3%. So we are closer to chimps than they are to gorillas.

Watson clarifies that the ancestor of all humans arose in Africa. And, interestingly, we can trace a common ancestor to all humans no more than 150,000 years ago.

Humans are almost genetically identical with each other. We are 99.9% alike. Variation among humans is very little compared with other species. Fruit flies are 10X more variant and chimps are 3X more variant. Why are humans so alike, Watson asks? He argues it is because our common ancestor was so recent, only 150,000 years ago, and this was "insufficient time for substantial variation to arise through mutation."

Watson makes the claim that under the fur of a chimp their skin is white. And that since we were connected to them genetically with a common ancestor 5 million years ago, black skin pigmentation later arose in human evolution to protect African skin from damaging sun rays (certainly necessary without as much body hair).

Though we are only 1% genetically different than chimps this is enough to account for enormous phenotype differences. For instance, we have language and they do not. The gene to account for why humans developed language and chimps did not is called the FOXP2 gene.

Near the book's end there is an important section on Edward O. Wilson's work on evolutionary psychology. Watson argues that Wilson's pivotal 1975 work should move from the "fringes of anthropology to the very heart of the discipline."

And, finally, Watson concludes his book was an argument for human genetic enhancement. But this idea has sometimes received a harsh reception, as it did when it was presented in Germany several years ago. Watson understands the ethical implication of this but hopes for a world with less disease and pain.

This reading reveals the wonderful awesomeness of DNA and a world so intimately connected through it. James Watson deserves applause for illuminating the "secret of life."

Nature Revealed: Selected Writings 1949-2006 *by Edward O. Wilson*

Edward O. Wilson compiled some of his classic work into this masterpiece. Highlighting the history of sociobiology, Wilson also spotlights a better integrated science known as "consilience." Both of these topics are indeed powerful tools to "reveal nature." Let us look at sociobiology and consilience separately:

Sociobiology:

Interestingly, Wilson's work began in 1949 with the study of fire ants. One question he asked was: Why is there a sterile caste in ants? He discovered that the unit of natural selection is the family unit and not the individual. Sterile ants work as helpers to serve reproductive interests of the family unit. Furthermore, his study of ants revealed that ants occupy the land for 100 million years. One form of social adaptation is slavery. Ants raid nests and make captives slaves.

Moreover, Wilson's work on ants led him to study pheromones. When pheromones where discovered in animals, Wilson's study of ants invisible odor served as strong evidence of these hormones. In 1963 there is the first notion of these hormones in humans.

In 1975 Wilson completed *Sociobiology: The New Synthesis*. While sociobiology was applied to animals in 1971, it was now applied to humans four years later. This new discipline eventually led to evolutionary psychology.

Evolutionary psychology shows that human decency / altruism arises from genes favoring them. Moral philosophy or ethics, as well as religion, can be explained with biology. Furthermore, Wilson explores the link between genetic evolution and cultural evolution….from genes to neural, mental development to mind to culture.

Wilson notes that there is a misunderstanding about sociobiology: it is not a belief that human behavior is genetically determined. But, according to Wilson, it is: "the study of the biological basis of all forms of social behavior in all organisms."

In 1980 Wilson focuses on global conservation movement and he becomes one of the fathers of this movement. One of his biggest fears is the loss of biological diversity as a result of humans. However, when discussing conservation ethics, he explains that biophilia is our natural affinity for life and our innately emotional affiliation with other organisms. Wilson points out that 99% of human history was hunter-gatherer bands and they were so intimately involved with life and organisms. Thus, the brain evolved, he argues, as biocentric.

Consilience:

Wilson's work on consilience work began in 1998. It is an attempt to connect natural science with social science/humanities. The term means "jumping together" of different fields of study. He refers to it as an "integrated science," where various disciplines are interlocked and reducible to the same general laws of nature.

The author dates the true origins of consilience with Thales of the 6th century BCE who appeared to be interested in a common network of explanations. With consilience, complex biological phenomena are reducible to the simple. For instance, biology can be explained in part by chemistry, which is in part can be explained by physics and then by quantum physics.

In today's age, there is now a connection between the natural science and the social sciences: neuroscience, psychology, philosophy are intertwined. Neural activity explains consciousness and consciousness explains culture. Thus, the mind has a "reducible material basis." Wilson argues that the social science will advance with consilience as it takes the reductionistic approach to human nature.

Now it can be argued that genetic evolution is linked directly to cultural evolution. Epigenetic rules determined by neurology / brain states established cross cultural universals (such as taboos against incest, fear of snakes, etc.)

Integrated science produces winners all around as it illustrates causally linked phenomena. There need not be distinct branches of learning but each is strengthened by assimilation. Overall, Wilson points out the obvious causal links between genes, mind and culture. As we are baffled by the meaning of our existence, an integrated understanding of the world is indeed necessary. Thanks to Wilson's seminal work, "nature is revealed" through this line of thinking.

The Descent of Man

Selected excerpts from Charles Darwin's Book

All that we know about savages, or may infer from their traditions and from old monuments, the history of which is quite forgotten by the present inhabitants, shew that from the remotest times successful tribes have supplanted other tribes. Relics of extinct or forgotten tribes have been discovered throughout the civilised regions of the earth, on the wild plains of America, and on the isolated islands in the Pacific Ocean. At the present day civilised nations are everywhere supplanting barbarous nations, excepting where the climate opposes a deadly barrier; and they succeed mainly, though not exclusively, through their arts, which are the products of the intellect. It is, therefore, highly probable that with mankind the intellectual faculties have been mainly and gradually perfected through natural selection; and this conclusion is sufficient for our purpose. Undoubtedly it would be interesting to trace the development of each separate faculty from the state in which it exists in the lower animals to that in which it exists in man; but neither my ability nor knowledge permits the attempt.

It deserves notice that, as soon as the progenitors of man became social (and this probably occurred at a very early period), the principle of imitation, and reason, and experience would have increased, and much modified the intellectual powers in a way, of which we see only traces in the lower animals. Apes are much given to imitation, as are the lowest savages; and the simple fact previously referred to, that after a time no animal can be caught in the same place by the same sort of trap, shews that animals learn by experience, and imitate the caution of others. Now, if some one man in a tribe, more sagacious than the others, invented a new snare or weapon, or other means of attack or defence, the plainest self-interest, without the assistance of much reasoning power, would prompt the other members to imitate him; and all would thus profit. The habitual practice of each new art must likewise in some slight degree strengthen the intellect. If the new invention were an important one, the tribe would increase in number, spread, and supplant other tribes. In a tribe thus rendered more numerous there would always be a rather greater chance of the birth of other superior and inventive members. If such men left children to inherit their mental superiority, the chance of the birth of still more ingenious members would be somewhat better, and in a very small tribe decidedly better. Even if they left no children, the tribe would still include their blood-relations; and it has been ascertained by agriculturists287 that by preserving and breeding from the family of an animal, which when slaughtered was found to be valuable, the desired character has been obtained.

Turning now to the social and moral faculties. In order that primeval men, or the apelike progenitors of man, should become social, they must have acquired the same instinctive feelings, which impel other animals to live in a body; and they no doubt exhibited the same general disposition. They would have felt uneasy when separated from their comrades, for whom they would have felt some degree of love; they would have warned each other of danger, and have given mutual aid in attack or defence. All this implies some degree of sympathy, fidelity, and courage. Such social qualities, the paramount importance of which to the lower animals is disputed by no one, were no doubt acquired by the progenitors of man in a similar manner, namely, through natural selection, aided by inherited habit. When two tribes of primeval man, living in the same country, came into competition, if (other circumstances being equal) the one tribe included a great number of courageous, sympathetic and faithful members, who were always ready to warn each other of danger, to aid and defend each other, this tribe would succeed better and conquer the other. Let it be borne in mind how all-important in the never-ceasing wars of savages, fidelity and courage must be. The advantage which disciplined soldiers have over undisciplined hordes follows chiefly from the confidence which each man feels in his comrades. Obedience, as Mr. Bagehot has well shewn, is of the highest value, for any form of government is better than none. Selfish and contentious people will not cohere, and without coherence nothing can be effected. A tribe rich in the above qualities would spread and be victorious over other tribes: but in the course of time it would, judging from all past history, be in its turn overcome by some other tribe still more highly endowed. Thus the social and moral qualities would tend slowly to advance and be diffused throughout the world.

But it may be asked, how within the limits of the same tribe did a large number of members first become endowed with these social and moral qualities, and how was the standard of excellence raised? It is extremely doubtful whether the offspring of the more sympathetic and benevolent parents, or of those who were the most faithful to their comrades, would be reared in greater numbers than the children of selfish and treacherous parents belonging to the same tribe. He who was ready to sacrifice his life, as many a savage has been, rather than betray his comrades, would often leave no offspring to inherit his noble nature. The bravest men, who were always willing to come to the front in war, and who freely risked their lives for others, would on an average perish in larger numbers than other men. Therefore, it hardly seems probable that the number of men gifted with such virtues, or that the standard of their excellence, could be increased through natural selection, that is, by the survival of the fittest; for we are not here speaking of one tribe being victorious over another.

Although the circumstances, leading to an increase in the number of those thus endowed within the same tribe, are too complex to be clearly followed out, we can trace some of the probable steps. In the first place, as the reasoning powers and foresight of the members became improved, each man would soon learn that if he aided his fellow-men, he would commonly receive aid in return. From this low motive he might acquire the habit of aiding his fellows; and the habit of performing benevolent actions certainly strengthens the feeling of sympathy which gives the first impulse to benevolent actions. Habits, moreover, followed during many generations probably tend to be inherited.

But another and much more powerful stimulus to the development of the social virtues, is afforded by the praise and the blame of our fellow-men. To the instinct of sympathy, as we have already seen, it is primarily due, that we habitually bestow both praises and blame on others, whilst we love the former and dread the latter when applied to ourselves; and this instinct no doubt was originally acquired, like all the other social instincts, through natural selection. At how early a period the progenitors of man in the course of their development, became capable of feeling and being impelled by, the praise or blame of their fellow-creatures, we cannot of course say. But it appears that even dogs appreciate encouragement, praise, and blame. The rudest savages feel the sentiment of glory, as they clearly show by preserving the trophies of their prowess, by their habit of excessive boasting, and even by the extreme care which they take of their personal appearance and decorations; for unless they regarded the opinion of their comrades, such habits would be senseless.

They certainly feel shame at the breach of some of their lesser rules, and apparently remorse, as shewn by the case of the Australian who grew thin and could not rest from having delayed to murder some other woman, so as to propitiate his dead wife's spirit. Though I have not met with any other recorded case, it is scarcely credible that a savage, who will sacrifice his life rather than betray his tribe, or one who will deliver himself up as a prisoner rather than break his parole, would not feel remorse in his inmost soul, if he had failed in a duty, which he held sacred.

We may therefore conclude that primeval man, at a very remote period, was influenced by the praise and blame of his fellows. It is obvious, that the members of the same tribe would approve of conduct which appeared to them to be for the general good, and would reprobate that which appeared evil. To do good unto others—to do unto others as ye would they should do unto you—is the foundation-stone of morality. It is, therefore, hardly possible to exaggerate the importance during rude times of the love of praise and the dread of blame. A man who was not impelled by any deep, instinctive feeling, to sacrifice his life for the good of others, yet was roused to such actions by a sense of glory, would by his example excite the same wish for glory in other men, and would strengthen by exercise the noble feeling of admiration. He might thus do far more good to his tribe than by begetting offspring with a tendency to inherit his own high character.

With increased experience and reason, man perceives the more remote consequences of his actions, and the self-regarding virtues, such as temperance, chastity, &c., which during early times are, as we have before seen, utterly disregarded, come to be highly esteemed or even held sacred. I need not, however, repeat what I have said on this head in the fourth chapter. Ultimately our moral sense or conscience becomes a highly complex sentiment—originating in the social instincts, largely guided by the approbation of our fellow-men, ruled by reason, self-interest, and in later times by deep religious feelings, and confirmed by instruction and habit.

It must not be forgotten that although a high standard of morality gives but a slight or no advantage to each individual man and his children over the other men of the same tribe, yet that an increase in the number of well-endowed men and an advancement in the standard of morality will certainly give an immense advantage to one tribe over another. A tribe including many members who, from possessing in a high degree the spirit of patriotism, fidelity, obedience, courage, and sympathy, were always ready to aid one another, and to sacrifice themselves for the common good, would be victorious over most other tribes; and this would be natural selection. At all times throughout the world tribes have supplanted other tribes; and as morality is one important element in their success, the standard of morality and the number of well-endowed men will thus everywhere tend to rise and increase.

It is, however, very difficult to form any judgment why one particular tribe and not another has been successful and has risen in the scale of civilisation. Many savages are in the same condition as when first discovered several centuries ago. As Mr. Bagehot has remarked, we are apt to look at the progress as normal in human society; but history refutes this. The ancients did not even entertain the idea, nor do the Oriental nations at the present day. According to another high authority, Sir Henry Maine, "The greatest part of mankind has never shewn a particle of desire that its civil institutions should be improved."290 Progress seems to depend on many concurrent favourable conditions, far too complex to be followed out. But it has often been remarked, that a cool climate, from leading to industry and to the various arts, has been highly favourable thereto. The Esquimaux, pressed by hard necessity, have succeeded in many ingenious inventions, but their climate has been too severe for continued progress. Nomadic habits, whether over wide plains, or through the dense forests of the tropics, or along the shores of the sea, have in every case been highly detrimental. Whilst observing the barbarous inhabitants of Tierra del Fuego, it struck me that the possession of some property, a fixed abode, and the union of many families under a chief, were the indispensable requisites for civilisation. Such habits almost necessitate the cultivation of the ground and the first steps in cultivation would probably result, as I have elsewhere shewn,291(2) from some such accident as the seeds of a fruit-tree falling on a heap of refuse, and producing an unusually fine variety. The problem, however, of the first advance of savages towards civilisation is at present much too difficult to be solved.

With savages, the weak in body or mind are soon eliminated; and those that survive commonly exhibit a vigorous state of health. We civilised men, on the other hand, do our utmost to check the process of elimination; we build asylums for the imbecile, the maimed, and the sick; we institute poor-laws; and our medical men exert their utmost skill to save the life of every one to the last moment. There is reason to believe that vaccination has preserved thousands, who from a weak constitution would formerly have succumbed to small-pox. Thus the weak members of civilised societies propagate their kind. No one who has attended to the breeding of domestic animals will doubt that this must be highly injurious to the race of man. It is surprising how soon a want of care, or care wrongly directed, leads to the degeneration of a domestic race; but excepting in the case of man himself, hardly any one is so ignorant as to allow his worst animals to breed.

The aid which we feel impelled to give to the helpless is mainly an incidental result of the instinct of sympathy, which was originally acquired as part of the social instincts, but subsequently rendered, in the manner previously indicated, more tender and more widely diffused. Nor could we check our sympathy, even at the urging of hard reason, without deterioration in the noblest part of our nature. The surgeon may harden himself whilst performing an operation, for he knows that he is acting for the good of his patient; but if we were intentionally to neglect the weak and helpless, it could only be for a contingent benefit, with an overwhelming present evil. We must therefore bear the undoubtedly bad effects of the weak surviving and propagating their kind; but there appears to be at least one check in steady action, namely that the weaker and inferior members of society do not marry so freely as the sound; and this check might be indefinitely increased by the weak in body or mind refraining from marriage, though this is more to be hoped for than expected.

In every country in which a large standing army is kept up, the finest young men are taken by the conscription or are enlisted. They are thus exposed to early death during war, are often tempted into vice, and are prevented from marrying during the prime of life. On the other hand the shorter and feebler men, with poor constitutions, are left at home, and consequently have a much better chance of marrying and propagating their kind.

Man accumulates property and bequeaths it to his children, so that the children of the rich have an advantage over the poor in the race for success, independently of bodily or mental superiority. On the other hand, the children of parents who are short-lived, and are therefore on an average deficient in health and vigour, come into their property sooner than other children, and will be likely to marry earlier, and leave a larger number of offspring to inherit their inferior constitutions. But the inheritance of property by itself is very far from an evil; for without the accumulation of capital the arts could not progress; and it is chiefly through their power that the civilised races have extended, and are now everywhere extending their range, so as to take the place of the lower races. Nor does the moderate accumulation of wealth interfere with the process of selection. When a poor man becomes moderately rich, his children enter trades or professions in which there is struggle enough, so that the able in body and mind succeed best. The presence of a body of well-instructed men, who have not to labour for their daily bread, is important to a degree which cannot be over-estimated; as all high intellectual work is carried on by them, and on such work, material progress of all kinds mainly depends, not to mention other and higher advantages. No doubt wealth when very great tends to convert men into useless drones, but their number is never large; and some degree of elimination here occurs, for we daily see rich men, who happen to be fools or profligate, squandering away their wealth.

Primogeniture with entailed estates is a more direct evil, though it may formerly have been a great advantage by the creation of a dominant class, and any government is better than none. Most eldest sons, though they may be weak in body or mind, marry, whilst the younger sons, however superior in these respects, do not so generally marry. Nor can worthless eldest sons with entailed estates squander their wealth. But here, as elsewhere, the relations of civilised life are so complex that some compensatory checks intervene. The men who are rich through primogeniture are able to select generation after generation the more beautiful and charming women; and these must generally be healthy in body and active in mind. The evil consequences, such as they may be, of the continued preservation of the same line of descent, without any selection, are checked by men of rank always wishing to increase their wealth and power; and this they effect by marrying heiresses. But the daughters of parents who have produced single children, are themselves, as Mr. Galton295 has shewn, apt to be sterile; and thus noble families are continually cut off in the direct line, and their wealth flows into some side channel; but unfortunately this channel is not determined by superiority of any kind.

Although civilisation thus checks in many ways the action of natural selection, it apparently favours the better development of the body, by means of good food and the freedom from occasional hardships. This may be inferred from civilised men having been found, wherever compared, to be physically stronger than savages. They appear also to have equal powers of endurance, as has been proved in many adventurous expeditions. Even the great luxury of the rich can be but little detrimental; for the expectation of life of our aristocracy, at all ages and of both sexes, is very little inferior to that of healthy English lives in the lower classes.

In regard to the moral qualities, some elimination of the worst dispositions is always in progress even in the most civilised nations. Malefactors are executed, or imprisoned for long periods, so that they cannot freely transmit their bad qualities. Melancholic and insane persons are confined, or commit suicide. Violent and quarrelsome men often come to a bloody end. The restless who will not follow any steady occupation—and this relic of barbarism is a great check to civilisation— emigrate to newly-settled countries; where they prove useful pioneers. Intemperance is so highly destructive, that the expectation of life of the intemperate, at the age of thirty for instance, is only 13.8 years; whilst for the rural labourers of England at the same age it is 40.59 years. Profligate women bear few children, and profligate men rarely marry; both suffer from disease. In the breeding of domestic animals, the elimination of those individuals, though few in number, which are in any marked manner inferior, is by no means an unimportant element towards success. This especially holds good with injurious characters which tend to reappear through reversion, such as blackness in sheep; and with mankind some of the worst dispositions, which occasionally without any assignable cause make their appearance in families, may perhaps be reversions to a savage state, from which we are not removed by very many generations. This view seems indeed recognised in the common expression that such men are the black sheep of the family.

Natural selection follows from the struggle for existence; and this from a rapid rate of increase. It is impossible not to regret bitterly, but whether wisely is another question, the rate at which man tends to increase; for this leads in barbarous tribes to infanticide and many other evils, and in civilised nations to abject poverty, celibacy, and to the late marriages of the prudent. But as man suffers from the same physical evils as the lower animals, he has no right to expect an immunity from the evils consequent on the struggle for existence. Had he not been subjected during primeval times to natural selection, assuredly he would never have attained to his present rank. Since we see in many parts of the world enormous areas of the most fertile land capable of supporting numerous happy homes, but peopled only by a few wandering savages, it might be argued that the struggle for existence had not been sufficiently severe to force man upwards to his highest standard. Judging from all that we know of man and the lower animals, there has always been sufficient variability in their intellectual and moral faculties, for a steady advance through natural selection. No doubt such advance demands many favourable concurrent circumstances; but it may well be doubted whether the most favourable would have sufficed, had not the rate of increase been rapid, and the consequent struggle for existence extremely severe. It even appears from what we see, for instance, in parts of S. America, that a people which may be called civilised, such as the Spanish settlers, is liable to become indolent and to retrograde, when the conditions of life are very easy. With highly civilised nations continued progress depends in a subordinate degree on natural selection; for such nations do not supplant and exterminate one another as do savage tribes. Nevertheless the more intelligent members within the same community will succeed better in the long run than the inferior, and leave a more numerous progeny, and this is a form of natural selection. The more efficient causes of progress seem to consist of a good education during youth whilst the brain is impressible, and of a high standard of excellence, inculcated by the ablest and best men, embodied in the laws, customs and traditions of the nation, and enforced by public opinion. It should, however, be borne in mind, that the enforcement of public opinion depends on our appreciation of the approbation and disapprobation of others; and this appreciation is founded on our sympathy, which it can hardly be doubted was originally developed through natural selection as one of the most important elements of the social instincts.

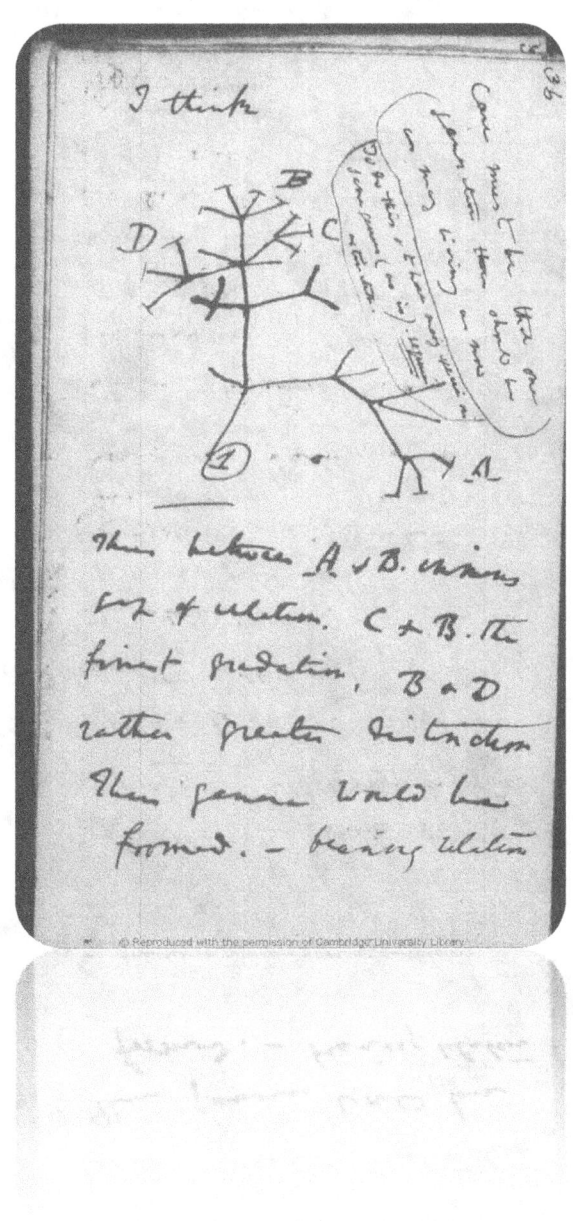

MSAC Philosophy Group
Mt. San Antonio College

DARWIN'S DNA:
A Brief Introduction to Evolutionary Philosophy

Dr. Andrea Diem-Lane

"Now turning our attention directly to philosophy we are in a better position to understand why the question "why" arises so often in human beings. In light of consciousness as a virtual simulator, any organism that can develop a mental "pivot" tool will have a tremendous advantage in thinking of new and unexpected strategies."

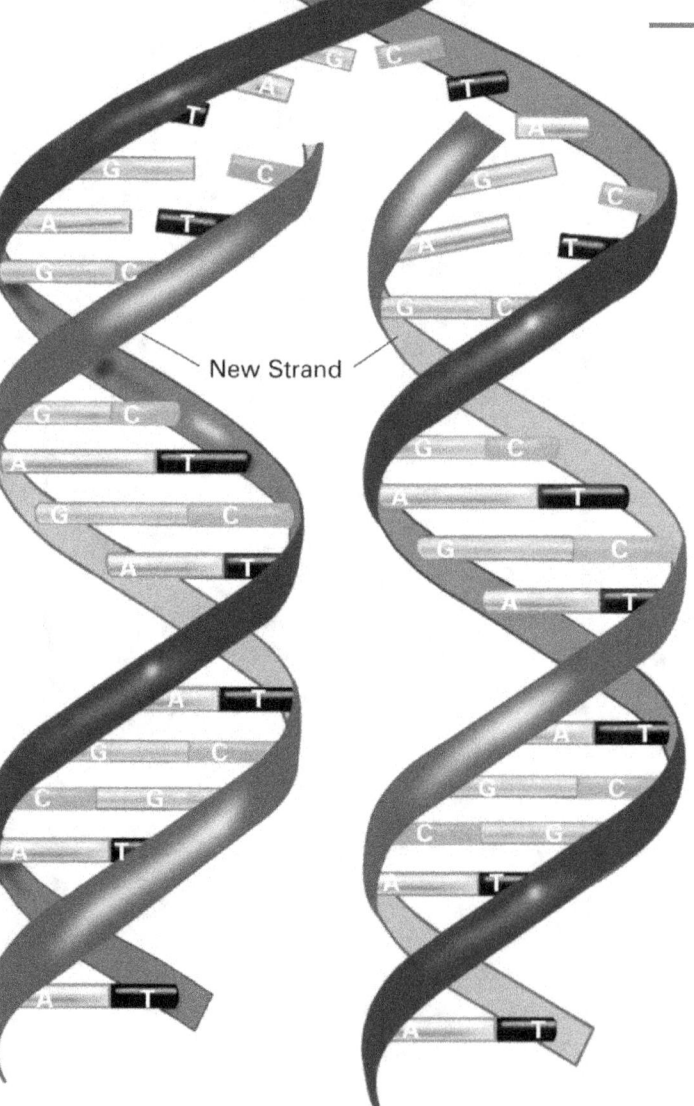

New Strand

ISBN: 978-1-56543-100-3

www.ingramcontent.com/pod-product-compliance
Lightning Source LLC
Chambersburg PA
CBHW080821170526
45158CB00009B/2494